Fundamentals of
ENVIRONMENTAL
ECONOMICS

The Authors

Dr. Arvind Kumar is the most senior faculty member of Post Graduate Department of Zoology, S.K.M. University, Dumka, Jharkhand and has 37 years experience as an outstanding teacher and researcher. He was also Pro-Vice Chancellor of Sido Kanhu Murma University, Dumka, Vice-Chancellor of Vinoba Bhave University, Hazaribagh (Jharkhand) and Vice-Chancellor, Magadh University, Bodh-Gaya (Bihar). Dr. Kumar is recipient of several awards and medals including Prof. W.A. Nizami Gold Medal by the Zoological Society of India, Dr. S.Z. Qasim Gold Medal by Bioved Research and Communication Centre, Allahabad.

Dr. I. Sundar, MA (Dev. St.), MA (Eco.), Ph.D., MA (Soc.) is working as Associate Professor (Economics) in Directorate of Distance Education, Annamalai University, Tamil Nadu. He has published more than 80 research papers in International and National Journals of repute. Besides this he has participated/delivered lectures at various National and International conferences/symposium etc.

Fundamentals of
ENVIRONMENTAL ECONOMICS

by
Professor A. Kumar
&
Dr. I. Sundar

2014
Daya Publishing House®
A Division of
Astral International Pvt. Ltd.
New Delhi – 110 002

Published by	:	**Daya Publishing House®**
		A Division of
		Astral International Pvt. Ltd.
		– ISO 9001:2008 Certified Company –
		4760-61/23, Ansari Road, Darya Ganj
		New Delhi-110 002
		Ph. 011-43549197, 23278134
		E-mail: info@astralint.com
		Website: www.astralint.com
Laser Typesetting	:	**Classic Computer Services**, Delhi - 110 035
Printed at	:	**Replika Press Pvt. Ltd.**

PRINTED IN INDIA

Preface

An understanding of economics is vital to any understanding of why environmental problems occur and what best to do about them.

Environmental economics is a relatively new field of economics that looks at environmental issues in relation to economic development and sustainability. Environmental economics looks a lot at environmental policies in countries, and how they impact the local and global economies, either positively or negatively. Environmental economics is generally viewed as a form of progressive economics, trying to account for various forms of market failures to better model markets in the future and lead to more widespread gains among people.

Society is driven by the uncontrollable movements of wealth, and Economists claim to be able to predict, influence, and control this primal force to achieve certain goals.

The conspiracy which has kept Economics alive and prospering works on several key principles.

Firstly, the whole field is broken down into thousands of macro- and micro-economic sub fields. When governments and corporations make use of an Economist's services, they must provide that Economist with information produced by others of that profession, which in turn is based on other information provided to the economist by the sponsoring organization. Therefore, any given economist whose results were proven to be false can blame the data provided to him by the organization he is working for. Since all the data is held in this loop and the original source cannot be determined, the government or corporation must either accept the blame for the failure, resulting in the loss of position at the highest level, or carry on as though the economic findings were in fact correct. In all cases, the latter path is followed and the economist gets a new car.

Secondly, economists have created several systems of government and political parties around the world. This shifts blame away from economists and toward the political party, who will tell the electorate, and then a new party will reign until the economists kill them too. Sometimes, however, economists will call for a regime change. If their suggestions are followed, initial failures of economic predictions will be blamed on the newness of the new regime. Later failures are blamed on mismanagement by the existing regime, and economists will call for an election, invasion, revolution or other change. The cycle may then begin again. And all the economists will get a new car.

Thirdly, their tenuous control over the flow of currency does allow economists to direct a large portion of the flow into their own bank accounts. This gives them power and leverage in many circumstances and has helped fund the international economic conspiracy. (That being, of course, that economists have way too damn many new cars.)

The present book is an accessible and up-to-date text covering all essential aspects of Environmental Economics. In this book, the readers can realize that, the environmental issues are viewed from an economic perspective, and therefore economic solutions to environmental protection are amply incorporated. The book comprises of fifteen chapters. The presentation of some of the chapters like the Externalities, Market Failure Analysis, Environment Kuznets Curve, Environmental Valuation, Natural Resource Valuation Techniques, Dose-Response Modelling, Cost–Benefit Analysis, Economics of Biodiversity, Global Warming, Acid Rain, Global Environmental Issues, Indian Environmental Challenges etc. is systematic, analytical, simple and lucid.

We hope that this book will be of greater use for the students, undergraduate courses, beginning graduate courses in environmental management and researchers to officials and policy makers, interested in improving the environmental quality.

Professor Arvind Kumar

Dr. I. Sundar

Contents

Preface   *v*

1. **Environmental Economics**   1

2. **Environmental Economics Basic Concepts**   24

3. **Externalities**   37

4. **Market Failure Analysis**   59

5. **Environment Kuznets Curve**   74

6. **Environmental Valuation**   86

7. **Natural Resource Valuation Techniques**   135

8. **Dose-Response Modelling**   155

9. **Cost–Benefit Analysis**   165

10. **Economics of Biodiversity**   179

11. **Global Warming**   196

12. **Acid Rain**   210

13. **Global Environmental Issues**   229

14. **Indian Environmental Challenges**   246

15. **Ability-to-Pay Principle**   264

Index   285

Chapter 1
Environmental Economics

Environmental economics is a relatively new field of economics that looks at environmental issues in relation to economic development and sustainability. Environmental economics looks a lot at environmental policies in countries, and how they impact the local and global economies, either positively or negatively. Environmental economics is generally viewed as a form of progressive economics, trying to account for various forms of market failures to better model markets in the future and lead to more widespread gains among people.

Society is driven by the uncontrollable movements of wealth, and Economists claim to be able to predict, influence, and control this primal force to achieve certain goals.

The conspiracy which has kept Economics alive and prospering works on several key principles.

Firstly, the whole field is broken down into thousands of macro- and micro-economic sub fields. When governments and corporations make use of an Economist's services, they must provide that Economist with information produced by others of that profession, which in turn is based on other information provided to the economist by the sponsoring organization. Therefore, any given economist whose results were proven to be false can blame the data provided to him by the organization he is working for. Since all the data is held in this loop and the original source cannot be determined, the government or corporation must either accept the blame for the failure, resulting in the loss of position at the highest level, or carry on as though the economic findings were in fact correct. In all cases, the latter path is followed and the economist gets a new car.

Secondly, economists have created several systems of government and political parties around the world. This shifts blame away from economists and toward the political party, who will tell the electorate, and then a new party will reign until the economists kill them too. Sometimes, however, economists will call for a regime change. If their suggestions are followed, initial failures of economic predictions will be blamed on the newness of the new regime. Later failures are blamed on mismanagement by the existing regime, and economists will call for an election, invasion, revolution or other change. The cycle may then begin again. And all the economists will get a new car.

Thirdly, their tenuous control over the flow of currency does allow economists to direct a large portion of the flow into their own bank accounts. This gives them power and leverage in many circumstances and has helped fund the international economic conspiracy. (That being, of course, that economists have way too damn many new cars.)

What is Environmental Economics?

A "standard" economic approach considers the economy as consisting of flows of goods and services between firms and households, influenced by the policy of governments.

Environmental economics adds to this the flow of resources from the environment (energy, raw materials, and so on) and the flow of waste products passing to the environment (various forms of pollution, solid waste, and so on).

Environmental economics has come into existence due to the realisation that humans can no longer ignore the effects they have on the environment through using resources and creating pollution. The "standard" model is fine for a world with low population, abundant resource, and no lasting environmental damage, but we no longer live in such a world.

More recently, a slightly different approach has arrived: ecological economics goes further than environmental economics, in recognising not only the flows between economy and environment, but also the complicated interactions and dynamics which can arise between human activity and ecosystems as a whole.

These topics are covered in detail in the course ENV 2A8Y. This week we're going to focus on Cost Benefit Analysis, which is a good introduction to some of the central ideas and controversies in environmental economics, and some basic resource management ideas, both in the context of forestry management.

Some Key Concepts

Economics is essentially concerned with the ways in which scarce resources are distributed among competing ends: if resources were not scarce, we would have no need for economics.

There are many views in ethics regarding how resources should be distributed among individuals. Any one of these can be incorporated within economic models. Indeed many economic models attempt to avoid making value judgments and limit themselves to describing what will happen if a certain policy is followed, and so on.

However some value judgments are widespread in economics. One of these is Individualism, which holds that welfare derives from the preferences of individuals. This is sometimes known as "consumer sovereignty". It implies no Paternalism and no concept of "social will".

Welfare is then measured by the concept of the utility of an individual: a measure of how "happy" an individual is. For many purposes, this doesn't have to be an absolute measure: it's enough to know if some change makes a person better off or worse off.

Another common value judgment is the Pareto principle, which states that a change is "good" if it makes some person better off, and makes nobody worse off.

This might seems "obvious", but in fact it is possible to argue against it–as a strict egalitarian would, for example. Nonetheless, the Pareto principle is widely accepted. There is one major problem with it, however, because it gives only a partial ranking of allocations: allocation "A" is preferred to "B" if and only if at least one person is better off in "A", and nobody is worse off; and vice versa. If some people are better off in "A" while others are better off in "B", then the Pareto principle cannot compare the two states of the world.

A common solution is that proposed by Kaldor and by Hicks: if the "winners" could fully compensate the "losers" and still be better off, then a change is "good". Note that the compensation is possible but need not actually take place. Ethically, this is much less steady ground. It is a short step from here to the idea of interpersonal comparisons of utility. If we can compare utility between different people, then we can combine the utilities of different people to give some measure of social welfare–this can be done in many ways, and depends heavily on value judgments.

Since an obvious goal for economic policy is to maximise social welfare, being able to measure this has obvious appeal for economists. However, the idea of comparing utility across individuals is very contentious. Despite the fact that many economists believe that utility cannot be compared between different people, the practice is widespread. In essence, this is for practical reasons: we do need some way in which to determine whether or not a change is a "good" thing, whether or not a policy should be introduced, or a project undertaken.

The way this tends to be done is to measure observable gains and losses in consumption. And in order to be able to add together consumption of different goods and services, monetary values are used. Very roughly, we measure an individual's welfare using the monetary value of that individual's consumption.

For marketed goods and services, the market price provides information on the value. But we are also interested in things which are not marketed: the health damage caused by pollution, for example, or the value of recreation in a forest.

Part of environmental economics is concerned with ways in which these non-marketed effects can be valued, in order to compare them with other, marketed goods and services. Clearly, valuing the environment is a contentious issue. For now, discuss in small groups for a couple of minutes the following questions:

☆ Why might we want to put a monetary "value" on human life?

☆ What range of values do you think might be used in practice?

☆ In what ways does your own behaviour reveal that your valuation of your own life is not infinite?

Environmental economics, which used to be on the periphery of the economics discipline, is fast becoming mainstream as concern for the environment grows. Practitioners in other disciplines (*e.g.* engineering, science, natural resource management, social sciences) are increasingly faced with environmental problems that have an economic component. This invaluable book fills an important gap in the literature by teaching both economists and non-economists how to use economic tools to address environmental problems.

Economics and the Environment

Economics may be defined as the 'study of choice under conditions of scarcity'. Historically, however, the environment has not generally been seen as a scarce resource; essentially the economic goods and services provided by the environment have been treated as having no cost. Robert Hahn summed it up thus: 'In earlier times, when environmental problems were few and far between, such ignorance was bliss and made good sense in some cases.' In recent decades society has begun to realise that 'the economy' and 'the environment' are fundamentally connected and that continued reliance on the environment as a limitless free source of natural resources or as a sink for wastes will not deliver sustainable development.

However, protection of the environment is not without cost. Every environmental protection measure has an opportunity cost, which is simply the foregone benefit of the measure, and is a fundamental tenet of economics. It can be illustrated by way of a simple example. A utility company laying a new pipeline can cut through a woodland or divert around it. The diversion, laying the pipe across agricultural land or in the public highway, would have increased cost due to the additional construction work, perhaps running into hundreds of thousands of pounds. It is appropriate to ask whether incurring this additional cost is worthwhile given that this expenditure could be invested in an alternative environmental protection measure, used to reduce customer charges, used to generate employment, or for some other beneficial purpose. As unpalatable as it may seem that such choices have to be made, some means of identifying the various costs and benefits of alternative project options is required to support decision makers aiming for sustainable development. Such analysis also allows them to make transparent and consistent trade-offs between economic, social and environmental objectives.

Environmental economics has emerged since the 1960s as a distinct branch of economics, although many of the essential principles can be traced far further back in time. Environmental economics revolves primarily around the failure of the market system to account for pollution and natural resource depletion. This failure manifests itself in the fact that developers are often not confronted with the full costs of their impact on the environment in market prices. In economic jargon, these impacts are known as externalities but as they affect third parties, it is important that they are taken into account. Essentially, environmental economics holds that the market failure needs to be addressed by 'getting the prices right': market prices should be corrected for external costs (or benefits) and, in response, developers will alter their consumption and/or production behaviour. For instance, over-abstraction of groundwater may have deleterious affects on wetland habitats and associated bird species. This may then have impacts on the level of recreation and tourism activities and on property values. The externality arises because the abstractor is not confronted with the costs of these impacts. Where such a market failure is corrected then, all other things being equal, increased costs will provide the incentive for reduced abstraction, for example by switching abstraction to a less damaging source.

Environmental economics is principally concerned with two aspects of such issues. Firstly the valuation of the externalities for subsequent use in the design of efficient policy measures and in project appraisal. Secondly the design of regulatory and incentive systems that include rewards and penalties that take these external values into account, *e.g.* pollution taxes and tradable permits. In both cases, the underlying framework is one of a formal analysis of costs and benefits. The objective is one of maximising net benefits to society, subject to other goals of public policy. Expressing environmental impacts in monetary terms allows a systematic comparison of costs and benefits. Such an approach allows a more rational judgement to be made of the balance between the costs and benefits of projects and policies, including environmental impacts.

Application of environmental economics can support the pursuit of sustainable development. Although there is no universally accepted definition of sustainable development, it is acknowledged that it involves balancing social, economic and environmental objectives of the current generation with those of future generations. This will unquestionably involve trade-offs between the different objectives over time. Environmental economics provides a robust framework for the illumination of these trade-offs, in particular between environmental and economic objectives.

Environmental Valuation

For conventional goods and services traded in markets identifying the costs and benefits is relatively straightforward as the prices of these goods and services are apparent. These prices signal the willingness to pay on the part of the buyer and the willingness to accept compensation on the part of the seller. However, because markets are absent for many of the goods and services provided by the environment, such as the capacities of rivers and the atmosphere to absorb pollution, prices cannot be directly observed. Therefore, in order to measure the costs and benefits of non-market environmental impacts environmental economists have developed a range of valuation techniques.

Revealed preference techniques allow the monetary value of the environmental impact to be derived through analysis of related markets. For example, all other things being equal, the prices of properties alongside polluted rivers will be lower than those adjacent to pristine rivers and the difference can be attributed to the value of the environmental impact. Given sufficient data, econometric models (known as hedonic pricing models) allow these values to be uncovered. In a similar vein, the money spent by individuals in visiting a recreation site may be used to infer the value of the site (this approach is known as the travel cost method). Revealed preference approaches have a major limitation however. By definition they can only capture use values, not the category of values known as non-use (or passive use) values. These values relate to the value that individuals hold for environmental goods and services independent of how much they actually use it. For instance, whilst econometric analysis can reveal the change in property values along a polluted watercourse, other individuals who do not live along, or visit, the watercourse may feel worse off just from knowledge of the pollution and its effects. All though non-use values are less tangible than use values they are nonetheless related to changes in individual well-being and it is therefore appropriate to include them in project and policy appraisal.

Because non-use values are, by definition, not reflected in any market environmental economists utilise stated preference techniques to measure them. The most common stated preference method is to conduct a 'contingent valuation' survey, where change in the level of provision of the environmental good or service in question is placed in a hypothetical market setting. This hypothetical scenario is presented to a sample of individuals from the affected population and they are invited to state their willingness to pay (or willingness to accept compensation) values as if they were trading in a conventional market. Statistical analysis of the survey responses allows environmental economists to calculate average willingness to pay values that can then be used to calculate the aggregate value for the entire population affected by the project or policy being assessed. Self-evidently, non-use values can extend over a wide population and thus frequently overshadow measures of use values in estimates of total economic value.

Environmental valuation is controversial. Critics question the ethics of placing monetary values on the environment from what they see as a purely selfish, human-centred motivation. The controversy could be lessened by understanding that it is not the environment itself that is being valued, rather individual preferences for the environment, being a measure of the well-being (or utility) that those affected attach to the goods or services provided by the environment. These preferences can be motivated by any number of factors, including altruism and concern for the rights of non-human species. Individuals may build into their valuation an element of intrinsic value, as they perceive an obligation on society to protect the environment for its own sake.

Even for those who accept environmental valuation and stated preference techniques, there are technical issues. These relate to eliciting individuals' true values using survey based methods as there are many biases that can influence responses. In the last decade there has been considerable progress in research in addressing these issues, prompted in a large part by the use of contingent valuation in estimating the environmental damage caused by the Exxon Valdez oil spill in Alaska in 1989. Following this event, a special commission established by the US government gave its qualified approval to use of contingent valuation in a legal context in the USA.

Because of the importance of non-use values contingent valuation and other stated preference methods have become the preferred approach to environmental valuation. Because of the relative youthfulness, and significant costs, of applying many of these techniques there are relatively few site-specific surveys that have been conducted in the UK. This leads to problems of accuracy and reliability as there is a temptation to transfer the results of one survey to another project or area, so called 'benefit transfer'. Credible benefit transfer requires sufficient similarity between the two populations and sites, but given the small database of values for transfer this is often difficult and often little consideration is given to differences between the populations. Benefit transfer has often been undertaken in situations where site-specific studies are warranted, often for reasons of cost, but this can be a false economy, as the understanding of benefits will be reduced. This situation will however improve with time as more site-specific surveys are completed and more research is undertaken.

The Key Issues

Environmental economics is a relatively new science, particularly in the UK, and we need to get more experience of its value in the decision making process. Expressing environmental costs and benefits in monetary terms is far from simple, and the methods at present in use can deliver a wide spread of values. This tends to discourage confidence in the methods. Standard approaches need to be developed where appropriate, and used consistently in order to produce more repeatable results, though realising that there can be valid reasons why values between apparently similar populations exhibit significant divergence. The UK government recognises this and has commissioned comprehensive guidance on the application of stated preference techniques. Ultimately, more site-specific surveys will add to our confidence in the validity and application of the techniques.

Environmental economics raises the fundamental question of who pays for environmental protection and how. Should costs be reflected in commodity prices, or should they be met through local or general taxation? These are issues that need to be debated.

An Economic View of the Environment

Economists are often charged with the difficult task of placing a value on environmental resources. The air we breathe and the water we drink are environmental resources that a re required to sustain life on this planet. The environment also serves as a source of pleasure to humans in viewing the sun on the horizon, admiring the tranquil beauty of a forest-ringed lake, or driving along a scenic ocean road. Many people argue that environmental (*e.g.*, land, air, and water) and natural resources (*e.g.*, coal, trees, and fish) are, in fact, priceless. Thus, they say, it is not appropriate to consider them within the context of economic valuation. However, the world's population has proven unwilling to forgo all future economic activity to eliminate all factors that cause environmental damage. Human societies are willing to give up certain environmental assets to generate other kinds of economic gains. The value placed on environmental assets is illustrated in our choices against other competing economic needs. Figure 1 shows an economic view of the environment and outlines the topics discussed below.

Economic principles suggest that the well-being of a society can be measured as the sum of all the individuals' level of well-being. This well-being, or what economists call "utility," is not derived solely from purchasing and consuming goods and services, but also from things like safety, and our physical, mental, and spiritual well-being.

The fact that all these things have utility is evident in that we are willing to trade our time, effort, money, and other resources to get them. The utility or degree of satisfaction experienced by individuals, and thus society, can be quantified in terms of the "willingness to pay" for goods and services, including environmental resources. In many cases, individuals do not pay for the environmental benefits they receive. However, their willingness to pay for these benefits can be derived from

An Economic View of the Environment
Measuring society's well-being
Willingness to pay as a measure of natural resource value
Classes of environmental values
Circular flow of the economy and environment
Market allocation of natural resources
Market failure
Government responses to market failure

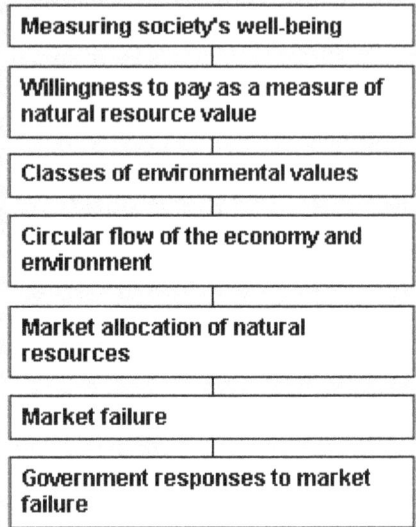

Figure 1: Overview of the topics discussed on this Web page.

surveys, observed behavior, or through other methods. The willingness-to-pay concept is key in environmental valuation, granting analysts a framework upon which to examine and measure individual preferences. Positive preferences for environmental resources translate into an expressed or observed willingness to pay for them. Conversely, individuals are not willing to pay for environmental resources that they do not value.

Coastal areas contain resource-rich environmental systems that provide a broad spectrum of services to humankind. From recreational opportunities, such as hiking and wildlife observation to the harvesting of fish and other seafood for human consumption, coastal habitats provide many direct benefits. Indirect benefits (*e.g.*, biological support and water and air purification) and non-use benefits (*e.g.*, the satisfaction that we get from knowing that the environment has been preserved for future generations) also affect the decision to exploit or conserve natural resources (Figure 2).

Natural resources are also valuable in the production of other goods. The output of any firm is a function of several important inputs, which economists call "factors of production." The factors of production include labour, capital (such as buildings and machinery), and an array of environmental inputs. These environmental inputs include the land upon which production takes place, raw materials extracted from the environment, such as minerals and timber, and often, clean air and water. It is sometimes, but not always, possible to offset declines in the natural resources required in production processes by increasing labour and capital. However, environmental depletion and damage ultimately lead to declining input availability and reductions of output and utility.

Figure 2: Direct benefits of the coastal environment include (a) whale watching and (b) shrimp harvesting, while an indirect benefit is (c) the habitat provided. Non-use benefits are derived from the satisfaction of knowing an area will be preserved for future generations (d).

The decisions society makes about how to use its natural and other resources involve tradeoffs. All other things being equal, when we decide to increase the flow of one service of coastal resources (*e.g.*, fish nursery), we are also implicitly deciding to decrease the flow of another service (*e.g.*, disposal of wastewater). In other words, our decision to harvest fish is tied to a decision to damage the fisheries output by disposing of polluted wastewater.

The decisions we make regarding resource use are governed by scarcity. To economists, scarcity is a function of supply and demand. If resources are highly valued, it is because they are scarce and the demands placed upon those resources are large relative to their availability. Further, the decision of how to allocate these scarce resources has an impact on all sectors of the economy because of the complex relationship between natural resource inputs and economic output.

The Circular Flow of the Economy and the Environment

The economy and the environment are inextricably linked. The environment supplies the raw materials and energy that are used to produce the goods that we consume. Waste generated by this production process is either recycled or dumped back into the environment. Figure 3, which represents a small part of the overall picture, partially demonstrates the interconnectivity of the economy and environment (Pearce and Turner 1990).

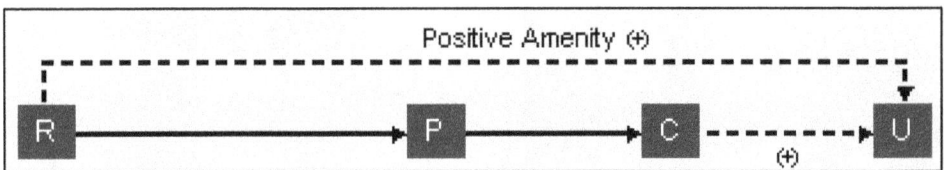

Figure 3: Resource inputs and positive amenity creates utility.

Later we will demonstrate that the relationship between the economy and environment does not represent an open, linear process, but is rather illustrative of a closed, circular system. Within the environment and economic flow diagram, raw materials (**R**) are used as inputs into the production process (**P**) that creates the goods consumed by households (**C**). The end result of production and consumption is the creation of utility (**U**) or satisfaction. Thus, the function of the environment, as highlighted within the diagram, is to provide material inputs into the production process and positive amenity to humankind.

In Figure 4, the resource box, **R**, is expanded to encompass two forms of natural resources: exhaustible and renewable resources (highlighted in red). **Exhaustible resources** (**ER**), which are not renewable, include oil, coal, and minerals. **Renewable resources** (**RR**), such as water and trees, may be replenished. In the diagram, "**h**" refers to the harvest of the resource and "**y**" to the sustainable yield. With respect to exhaustible resources, the harvest always exceeds the sustainable yield because these resources have no regenerative capacity. For renewable resources, the resource stock will decline if the harvest exceeds the yield but may actually grow if the harvest is less than the environment's capacity for regeneration. Thus, to guarantee the continued use of a renewable resource, it must be harvested at a rate slower than the sustainable yield.

The diagram is expanded in Figure 5 to include the generation of waste products (**W**) (new elements are highlighted in green). Waste products arise from processing or mining of resources. Waste products, such as the emissions and solid waste generated by industrial facilities, are also created by the production process. Final consumers

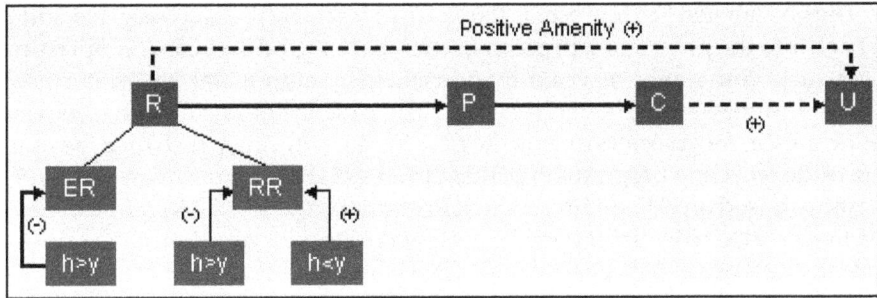

Figure 4: Exhaustible and renewable resources.

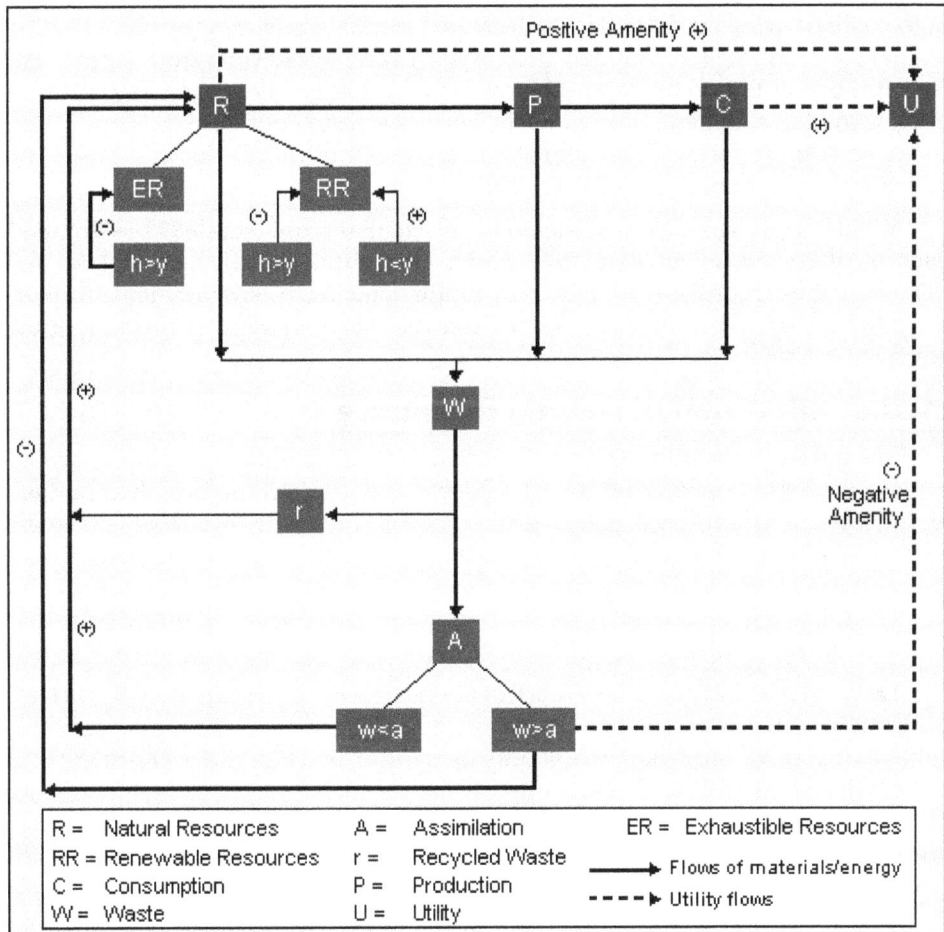

R =	Natural Resources	A =	Assimilation	ER = Exhaustible Resources
RR =	Renewable Resources	r =	Recycled Waste	──────► Flows of materials/energy
C =	Consumption	P =	Production	- - - ► Utility flows
W =	Waste	U =	Utility	

Figure 5: Interconnectivity of the economy and environment.

also create waste by disposing of product packaging and, ultimately, the product itself.

The box labeled "**r**" represents the share of total waste that is recycled and thus put back into the production process. Bottles, paper products, cans and plastics are all products commonly recycled by households. Scrap metal and water used in industrial processing are often recycled. The practice of recycling has expanded significantly in most sectors during the past 20 years, during which time the number of curbside recycling programs expanded from a single program to more than 9,000 programs. Based on U.S. Environmental Protection Agency (EPA) 2001 estimates of recycling rates for consumer goods, newspapers are now recycled at the rate of 60.2 per cent, steel cans at 58.1 per cent and glass containers at 22.0 per cent (EPA 2004). Note, however, new practices in sectors such as farming are running counter to those trends and have implications for coastal management. For example, the expansion of concentrated animal feeding operations or CAFOs (*i.e.*, greater than 1,000 animal units confined at a single location) has decreased recycling in farming by concentrating animal waste in smaller, confined areas. Manure and wastewater from CAFOs can contribute to pollution by depositing excessive amounts of nitrogen, phosphorus, organic matter, sediment, heavy metals, hormones, and antibiotics in streams and rivers adjacent to farms. EPA estimates the number of CAFOs located across the United States at roughly 15,000 (EPA 2003).

What happens to the remainder of the waste that is not recycled? It is dumped back into the environment. From the industrial and municipal sewage that flows into the seas to the carbon dioxide emitted by motor vehicles, the environment serves as the ultimate repository of many waste products. Thus, the second function of the environment is to serve as a waste sink (**W**).

Market Allocation of Natural Resources

The goals of consumers and producers are in conflict. Rational consumers try to achieve the highest level of utility that is possible within the limits of their budget, and rational producers try to maximize their profits. Lower prices enable consumers to purchase more of a good, thus expanding their utility. However, lower prices reduce the revenues, and thus profits, that accrue to producers. In a market economy, these conflicting goals are reconciled at a competitive market equilibrium price that balances the forces of supply and demand.

Demand is a schedule of how much of a good or service individuals will purchase during a specified period, depending on price and other factors. The law of demand states that as prices increase, the quantity demanded will fall, and as the price falls, more will be demanded, all other things being equal. Supply is a schedule of how much of a good or service firms supply during a specified period, depending on price and other factors. The law of supply states that as prices grow, the quantity supplied will increase, and as prices fall, firms supply less to the market.

Figure 6 demonstrates how demand and supply work together to determine the price of a commodity. Other factors being unchanged, the demand curve shows the relationship between price and quantity demanded, whereas the supply curve shows the relationship between price and the quantity supplied. Figure 6 demonstrates these concepts by examining the market for farmed oysters. In Figure 6, the competitive

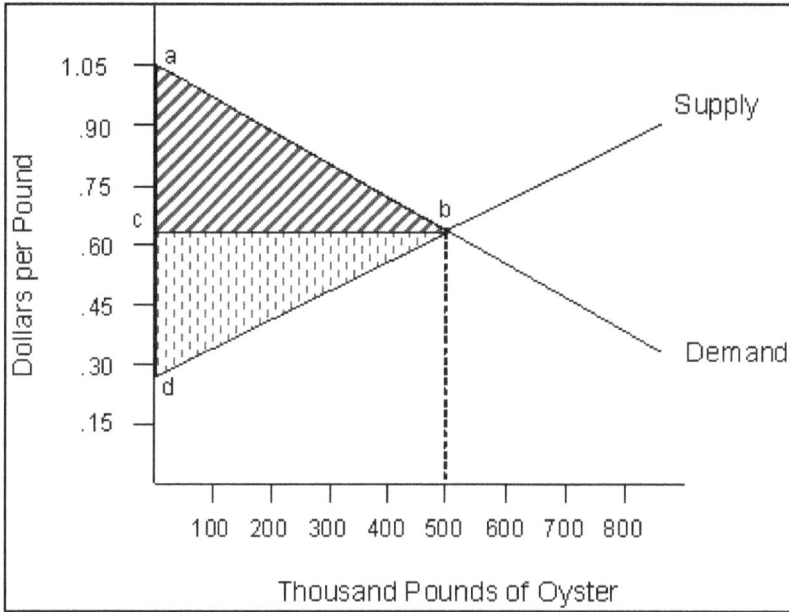

Figure 6: Supply and demand for farmed oysters.

market equilibrium (point b) is reached at a price of $0.62 per pound of oysters, resulting in a supply of 500 thousand pounds. Note, however, that unlike oyster farms, most fisheries are characterized as common property resources with open access, and fisheries often become overfished economically and sometimes biologically as well. For example, the market supply curve for salmon does not represent the marginal cost of supplying salmon because of the common property aspect of the resource combined with open access to the fishers. These two institutional characteristics lead to application of additional capital and labor until economic returns are equal to average (rather than marginal) costs and the economic benefits of fishing are dissipated. When a fishery is being overfished from a biological standpoint, production will not be sustainable. Thus, any benefits associated with habitat restoration could be negated by overfishing. There are techniques, however, to resolve the problem of overfishing, including limited entry programmes and individual transferable quotas (ITQs). These institutional solutions must also be considered when conducting habitat restoration.

The supply curve for farmed oysters mirrors the marginal cost to the oyster farming industry of harvesting oysters (*e.g.*, purchasing juvenile oysters, purchasing algae for feeding, growth monitoring, extraction, and transportation). Provided that the full costs to society associated with harvesting oysters (including environmental) are captured, the equilibrium point "b" highlighted in Figure 3 will result in an efficient resource allocation and a maximum level of economic benefit to society. At any quantity below 500 thousand pounds of oysters, the benefits associated with increasing the yield would exceed the costs of harvesting the oysters. Alternatively, any quantity that exceeds the market equilibrium would result in costs to industries that exceed the benefits supplied to the consumer.

The supply and demand for farmed oysters, however, are not static and can shift over time. A number of variables can lead to a shift in the entire demand curve for a product. These variables include tastes and preferences, the number of buyers, income, prices of substitute goods, prices of complement goods, and expectations. Figure 7 demonstrates how a change in one of these shift variables can lead to an increase in demand. Imagine that a medical report hailed the health benefits of oyster consumption. The demand for oysters would increase and a new equilibrium price ($0.76) and quantity (680 thousand pounds) would be reached. A number of variables can cause a shift in supply as well, including resource prices, the number of sellers, technology changes, prices of alternative outputs, expectations, taxes, and subsidies.

Economists measure the net economic benefit in a market as the difference between what it costs to produce a good or service, on the one hand, and what consumers are willing to pay for it, on the other. In an efficiently functioning competitive market, the net economic benefit is divided between consumers and producers. The net economic benefit is, therefore, divided between what is known as consumer and producer surplus. Consumer surplus is the difference between what each customer is willing to pay at each point in time and the price of the good or service and is represented by the area falling above the price line and below the demand curve. Consumer surplus in Figure 6 is represented by area *abc* and is represented by area *gef* in Figure 7.

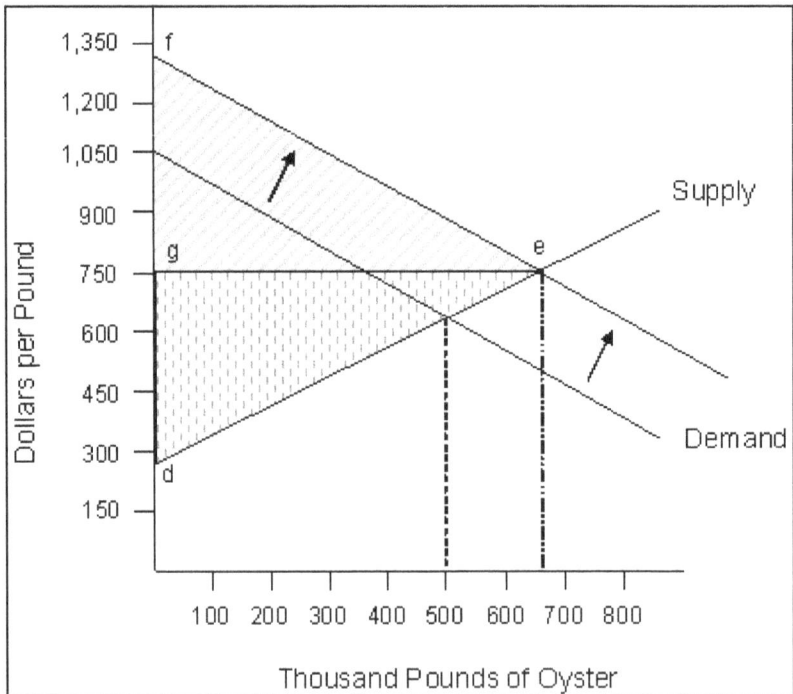

Figure 7: Shifting demand for farmed oysters.

Producer surplus is the difference between what a supplier is paid for a good or service and what it costs to supply, and is represented by area *bcd* in Figure 6 and area *deg* in Figure 7. The total economic benefit of a sale is the sum of the consumer and producer surplus. Consumer and producer surplus is a function of both supply and demand. Figure 8 demonstrates that as demand increases as represented in the outward shift in the demand curve, consumer and producer surplus is increased.

Market Failure

The market represents a decentralized exchange mechanism that enables society to allocate resources efficiently. Markets, however, efficiently can fail to allocate environmental assets through the price mechanism if they are unable to accurately capture the full social costs of exploiting the natural resource. Thus, there are a number of factors (*e.g.*, imperfect information, uncompensated environmental damage) that constrain the capacity of the market to achieve an accurate competitive market equilibrium. For example, the previous analysis would not measure net economic benefit if the oyster industry damaged the environment while harvesting. These are additional societal costs that are not necessarily reflected in the industry supply curve and market clearing price. When there are factors that prevent the market from achieving an efficient allocation of resources, market failure is evident. Externalities and nonprivate goods are two of the most recognized forms of market failure.

Many economic activities may provide secondary benefits or impose spillover costs to individuals and to society. These secondary effects, which are not recognized in the market transaction, are referred to as externalities. A negative externality occurs when the byproduct of an economic activity imposes a cost on society not captured in the market. For instance, motor vehicle emissions (*e.g.*, carbon dioxide, nitrogen oxide) contribute to the warming of the troposphere through the greenhouse effect. In turn, atmospheric warming contributes to rising sea levels. Rising seas, in turn, lead to increased flooding and corresponding loss of coastal wetlands. These costs are not captured through registration fees or the price of a gallon of gasoline.

Returning to our example, the full social cost of harvesting oysters is captured by the social supply curve in Figure 8, reflecting the marginal cost to society of consuming oysters. Because the firm does not pay the full economic or opportunity cost of providing the good, the private supply curve is too low and the good will be oversupplied (500 thousand pounds of oysters rather than 380 thousand pounds) and offered at a price below the social optimum ($0.62 rather than $0.75). Thus, the benefits of consuming oysters are exceeded by the full social costs of harvesting the resource. The loss of benefits due to overharvesting is illustrated in area *abc*.

Non-private goods represent another form of market failure. Most goods in our economy are classified according to two criteria: excludability and rivalness. Excludability is present when ownership is clearly identified and benefits accrue only to the owner. Rivalness is present when the owner's capacity to derive utility by consuming the good or service diminishes the capacity of others to enjoy the same benefit. Table 1 presents a typology of goods. The four types of goods highlighted are (*a*) private goods, (*b*) common resources, (*c*) club goods, and (*d*) public goods.

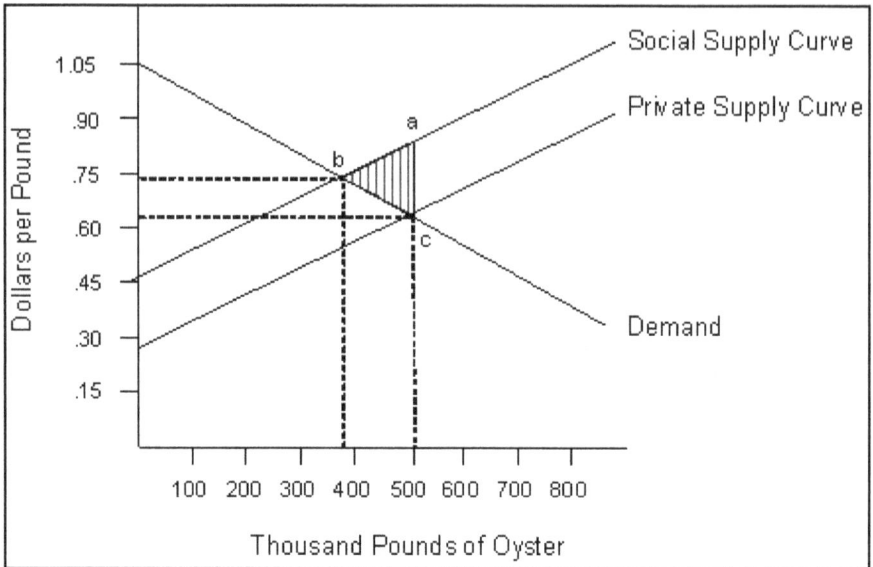

Figure 8: Market failure.

How do public goods vary from private goods? Private goods are exchanged in a market with buyers and sellers agreeing on a price and exchanging ownership rights. Private goods are, therefore, excludable, meaning that owners are clearly identified and the stream of benefits accrue only to the owner. Private goods also demonstrate rivalness.

Table 1: Typology of goods.

		Rival?	
		Yes	**No**
Excludable?	**Yes**	**Private Goods** - Clothing - Housing - Automobiles	**Club Goods** - Fire protection - Cable television - Electrical power
	No	**Common Resources** - Fish in the ocean - Part of the environment - Congested public roads	**Public Goods** - Un-congested public roads - Air - Oceans

Public goods on the other hand are nonexcludable and non-rival in competition. Oceans provide a good example. An ocean will not cease to exist, even if large numbers of individuals enjoy its benefits. Further, one consumer generally cannot prevent another from enjoying the recreational and amenity values an ocean provides, nor can they claim ownership of an ocean. Other examples of public goods include clean air and public roadways. In turn, externalities arise from public goods because there are no prices attached to a good that has value. Thus, individuals receive benefits without paying for them.

Public goods are at one end of the spectrum (nonrival, non-exludable), completely private goods (rival, excludable) at the other end, and goods with varying degrees of excludability and rivalness are located in between. **Common resources** are not excludable but are rival goods because one person's use of the common resource reduces the benefits that accrue to other users. Thus these resources are available to any consumer at no cost, but consumption of the resources diminishes their availability to other consumers. Common resources tend to be overexploited or used excessively. The government can solve the problems associated with common resources by defining property rights and by regulating private behavior (*e.g.*, catch limits, taxes). **Club goods** are excludable but nonrival. Examples of club goods include toll roads, swimming pools, fire protection, satellite television transmission, and electrical power. Environmental regulators are using the concept of excludability to regulate and protect natural resources. The opportunity to regulate a club good occurs when the government exercises its power as a "gatekeeper" and restricts entry to companies that pay into an externality-remediation fund or adopt practices that limit externalities.

Market failure is evident in the case of many environmental resources, including those associated with coastal habitat. For example, sport fishing and wildlife viewing provide an important source of recreation and amenity to the population. When private firms damage the quality of the habitat and reduce the recreational and amenity value it provides, governments often intervene to reduce the harmful impact of the environmental externality. For example, water quality permits are issued by states to individuals and businesses that discharge pollutants into surface and ground waters. Water quality permits are issued to protect surface and ground waters by regulating sewage and wastewater discharges and stormwater runoff from industrial and construction-related activities. Discharges can occur through a number of sources, including irrigation, on-site sewage systems, dry wells, and seepage ponds. The permit, which generally varies in price based on the type of operation and anticipated discharge levels, is designed to monetize and collect compensation for the impact of an environmental externality (*i.e.*, decline in water quality) resulting from construction and industrial activities.

The fundamental economic principles detailed in this article provide the framework for the valuation of environmental resources, a topic covered in another portion of this Web site. Valuation of environmental resources enables planners and policymakers to weigh environmental policies and strategies and select the course of action that yields the most benefits to society.

This introductory article focuses on the following main points:

☆ Economists measure the well-being of a society as the sum of all the individuals' well-being, or what economists call "utility."

☆ The level of utility or degree of satisfaction experienced by individuals, and thus society, can be quantified in terms of the "willingness to pay" for goods and services, including environmental resources. The willingness-to-pay concept is key in environmental valuation, granting analysts a framework upon which to examine and measure individual preferences.

☆ Coastal areas provide a broad range of values to humankind. Environmental values are categorized as direct use (*e.g.*, hiking, wildlife observation, inputs into the production process), indirect (*e.g.*, biological support, water and air purification), and nonuse (bequest and existence values). Each of these values affects the decision to exploit or conserve natural resources.

☆ The environment and economy are inextricably linked. The environment supplies the raw materials and energy that are used to produce the goods that we consume. Waste generated by the production process is either recycled or dumped back into the environment. Thus, the relationship between the environment and the economy does not represent an open, linear process, but is rather illustrative of a closed, circular system.

☆ The demand for products and the supply or availability of natural resources determine the market allocation of natural resources. A number of factors (*e.g.*, scarcity, tastes and preferences, the number of buyers, income, prices of substitute goods, prices of complement goods, expectations, resource prices, number of sellers, technology change, prices of alternative outputs, and taxes and subsidies) collectively determine the supply and demand for products, thus dictating how best to allocate scarce natural resources efficiently.

☆ Markets can fail to allocate natural resources efficiently through the pricing mechanism if they are unable accurately to capture the full social costs of exploiting the natural resource. When there are factors that prevent the market from achieving an efficient allocation of resources, market failure is evident.

Economists agree that markets are generally most well-suited for efficiently allocating society's resources. Markets can fail, however, resulting in the byproduct on an economic activity, generally referred to as an externality, to impose costs on society not captured in the market. To combat environmental externalities and other forms of market failure, government agencies use various strategies (*e.g.*, permits, restrictions to access, regulations, equipment restrictions, taxes) designed to correct these market imperfections and protect the environment. Finding an appropriate balance between the demands of the marketplace and the need to preserve natural resources enables our society both to expand the economy and preserve the environment for future generations.

Topics and Concepts

Central to environmental economics is the concept of market failure. Market failure means that markets fail to allocate resources efficiently. As stated by Hanley, Shogren, and White (2007) in their textbook *Environmental Economics*: "A market failure occurs when the market does not allocate scarce resources to generate the greatest social welfare. A wedge exists between what a private person does given market prices and what society might want him or her to do to protect the environment. Such a wedge implies wastefulness or economic inefficiency; resources can be reallocated to make at least one person better off without making anyone else worse off." Common forms of market failure include externalities, non excludability and non rivalry.

Externality: the basic idea is that an externality exists when a person makes a choice that affects other people that are not accounted for in the market price. For instance, a firm emitting pollution will typically not take into account the costs that its pollution imposes on others. As a result, pollution in excess of the 'socially efficient' level may occur. A classic definition is provided by Kenneth Arrow (1969), who defines an externality as "a situation in which a private economy lacks sufficient incentives to create a potential market in some good, and the non-existence of this market results in the loss of efficiency." In economic terminology, externalities are examples of market failures, in which the unfettered market does not lead to an efficient outcome.

Common property and non-exclusion: When it is too costly to exclude people from accessing a rivalrous environmental resource, market allocation is likely to be inefficient. The challenges related with common property and non-exclusion have long been recognized. Hardin's (1968) concept of the tragedy of the commons popularized the challenges involved in non-exclusion and common property. "commons" refers to the environmental asset itself, "common property resource" or "common pool resource" refers to a property right regime that allows for some collective body to devise schemes to exclude others, thereby allowing the capture of future benefit streams; and "open-access" implies no ownership in the sense that property everyone owns nobody owns. The basic problem is that if people ignore the scarcity value of the commons, they can end up expending too much effort, over harvesting a resource (*e.g.*, a fishery). Hardin theorizes that in the absence of restrictions, users of an open-access resource will use it more than if they had to pay for it and had exclusive rights, leading to environmental degradation. See, however, Ostrom's (1990) work on how people using real common property resources have worked to establish self-governing rules to reduce the risk of the tragedy of the commons.

Public goods and non-rivalry: Public goods are another type of market failure, in which the market price does not capture the social benefits of its provision. For example, protection from the risks of climate change is a public good since its provision is both non-rival and non-excludable. Non-rival means climate protection provided to one country does not reduce the level of protection to another country; non-excludable means it is too costly to exclude any one from receiving climate protection. A country's incentive to invest in carbon abatement is reduced because it can "free ride" off the efforts of other countries. Over a century ago, Swedish economist Knut Wicksell (1896) first discussed how public goods can be under-provided by the market because people might conceal their preferences for the good, but still enjoy the benefits without paying for them.

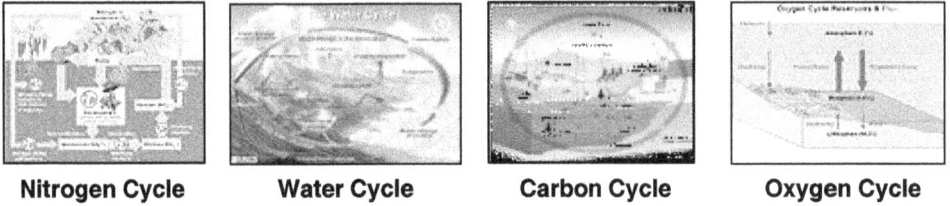

| Nitrogen Cycle | Water Cycle | Carbon Cycle | Oxygen Cycle |

Figure 9: Global geochemical cycles critical for life.

Valuation

Assessing the economic value of the environment is a major topic within the field. Use and indirect use are tangible benefits accruing from natural resources or ecosystem services (see the nature section of ecological economics). Non-use values include existence, option, and bequest values. For example, some people may value the existence of a diverse set of species, regardless of the effect of the loss of a species on ecosystem services. The existence of these species may have an option value, as there may be possibility of using it for some human purpose (certain plants may be researched for drugs). Individuals may value the ability to leave a pristine environment to their children.

Use and indirect use values can often be inferred from revealed behaviour, such as the cost of taking recreational trips or using hedonic methods in which values are estimated based on observed prices. Non-use values are usually estimated using stated preference methods such as contingent valuation or choice modelling.

Solutions

Solutions advocated to correct such externalities include:

Environmental Regulations

Under this plan the economic impact has to be estimated by the regulator. Usually this is done using cost-benefit analysis. There is a growing realization that regulations (also known as "command and control" instruments) are not so distinct from economic instruments as is commonly asserted by proponents of environmental economics. For example, 1 regulations are enforced by fines, which operate as a form of tax if pollution rises above the threshold prescribed. 2 pollution must be monitored and laws enforced, whether under a pollution tax regime or a regulatory regime. The main difference an environmental economist would argue exists between the two methods, however, is the total cost of the regulation. "Command and control" regulation often applies uniform emissions limits on polluters, even though each firm has different costs for emissions reductions. Some firms, in this system, can abate inexpensively, while others can only abate at high cost. Because of this, the total abatement has some expensive and some inexpensive efforts to abate. Environmental economic regulations find the cheapest emission abatement efforts first, then the more expensive methods second. For example, as said earlier, trading, in the quota system, means a firm only abates if doing so would cost less than paying someone else to make the same reduction. This leads to a lower cost for the total abatement effort as a whole.

Quotas on Pollution

Often it is advocated that pollution reductions should be achieved by way of tradeable emissions permits, which if freely traded may ensure that reductions in pollution are achieved at least cost. In theory, if such tradeable quotas are allowed, then a firm would reduce its own pollution load only if doing so would cost less than paying someone else to make the same reduction. In practice, tradeable permits approaches have had some success, such as the U.S.'s sulphur dioxide trading program, though interest in its application is spreading to other environmental problems.

Taxes and Tariffs on Pollution/Removal of "Dirty Subsidies"

Increasing the costs of polluting will discourage polluting, and will provide a "dynamic incentive", that is, the disincentive continues to operate even as pollution levels fall. A pollution tax that reduces pollution to the socially "optimal" level would be set at such a level that pollution occurs only if the benefits to society (for example, in form of greater production) exceeds the costs. Some advocate a major shift from taxation from income and sales taxes to tax on pollution–the so-called "green tax shift".

Better Defined Property Rights

The Coase Theorem states that assigning property rights will lead to an optimal solution, regardless of who receives them, if transaction costs are trivial and the number of parties negotiating is limited. For example, if people living near a factory had a right to clean air and water, or the factory had the right to pollute, then either the factory could pay those affected by the pollution or the people could pay the factory not to pollute. Or, citizens could take action themselves as they would if other property rights were violated. The US River Keepers Law of the 1880s was an early example, giving citizens downstream the right to end pollution upstream themselves if government itself did not act (an early example of bioregional democracy). Many markets for "pollution rights" have been created in the late twentieth century – see emissions trading. The assertion that defining property rights is a solution is controversial within the field of environmental economics and environmental law and policy more broadly; in Anglo-American and many other legal systems, one has the right to carry out any action unless the law expressly proscribes it. Thus property rights are already assigned (the factory that is polluting has a right to pollute).

Relationship to other Fields

Environmental economics is related to ecological economics but there are differences. Most environmental economists have been trained as economists. They apply the tools of economics to address environmental problems, many of which are related to so-called market failures – circumstances wherein the "invisible hand" of economics is unreliable. Most ecological economists have been trained as ecologists, but have expanded the scope of their work to consider the impacts of humans and their economic activity on ecological systems and services, and vice-versa. This field takes as its premise that economics is a strict subfield of ecology. Ecological economics is sometimes described as taking a more pluralistic approach to environmental problems and focuses more explicitly on long-term environmental sustainability and issues of scale.

These two groups of specialists sometimes have conflicting views which can often be traced to the different philosophical underpinnings of the two fields. Some ecologists subscribe to deontological ethical systems; other economists subscribe to teleological ethical systems. Neither ethical system can be demonstrated to be right or wrong, but they may sometimes have different implications for environmental policy. Environmental economics is viewed as relatively more pragmatic in a price system; ecological economics as relatively more idealistic as it supposedly does not use money to arbiter decision making as much.

Another context in which externalities apply is when globalization permits one player in a market who is unconcerned with biodiversity to undercut prices of another who is – creating a "race to the bottom" in regulations and conservation. This in turn may cause loss of natural capital with consequent erosion, water purity problems, diseases, desertification, and other outcomes which are not efficient in an economic sense. This concern is related to the subfield of sustainable development and its political relation, the anti-globalization movement.

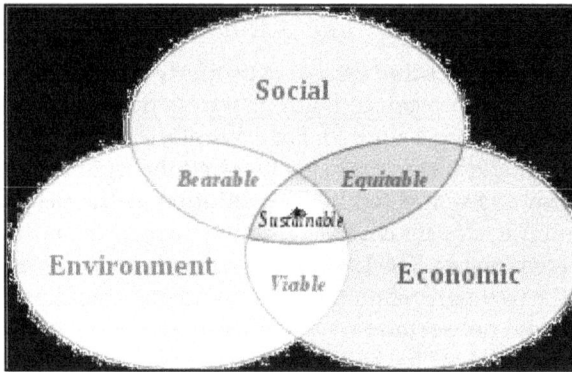

Figure 10

The Three Pillars of Sustainability

Environmental economics was once distinct from resource economics. Natural resource economics as a subfield began when the main concern of researchers was the optimal commercial exploitation of natural resource stocks. But resource managers and policy-makers eventually began to pay attention to the broader importance of natural resources (*e.g.*, values of fish and trees beyond just their commercial exploitation;, externalities associated with mining). It is now difficult to distinguish "environmental" and "natural resource" economics as separate fields as the two became associated with sustainability. Many of the more radical green economists split off to work on an alternate political economy.

Environmental economics was a major influence for the theories of natural capitalism and environmental finance, which could be said to be two sub-branches of environmental economics concerned with resource conservation in production, and the value of biodiversity to humans, respectively. The theory of natural capitalism (Hawken, Lovins, Lovins) goes further than traditional environmental economics by envisioning a world where natural services are considered on par with physical capital.

The more radical Green economists reject neoclassical economics in favour of a new political economy beyond capitalism or communism that gives a greater emphasis to the interaction of the human economy and the natural environment, acknowledging that "economy is three-fifths of ecology"– Mike Nickerson.

These more radical approaches would imply changes to money supply and likely also a bioregional democracy so that political, economic, and ecological "environmental limits" were all aligned, and not subject to the arbitrage normally possible under capitalism.

Chapter 2
Environmental Economics Basic Concepts

The Law of Diminishing Returns

The "law of diminishing returns" is one of the best-known principles outside the field of economics. It was first developed in 1767 by the French economist Turgot in relation to agricultural production, but it is most often associated with Thomas Malthus and David Ricardo. They believed that human population would eventually outpace the production of food since land was an integral factor in limited supply. In order to increase production to feed the population, farmers would have to use less fertile land and/or increase production intensity on land currently under production. In both cases, there would be diminishing returns.

The law of diminishing returns – which is related to the concept of marginal return or marginal benefit – states that if one factor of production is increased while the others remain constant, the marginal benefits will decline and, after a certain point, overall production will also decline. While initially there may be an increase in production as more of the variable factor is used, eventually it will suffer diminishing returns as more and more of the variable factor is applied to the same level of fixed factors, increasing the costs in order to get the same output. Diminishing returns reflect the point in which the marginal benefit begins to decline for a given production process. For example, the Table 1 sets the conditions on a farm producing corn.

It is with three workers that the farm production is most efficient because the marginal benefit is at its highest. Beyond this point, the farm begins to experience diminishing returns and, at the level of 6 workers, the farm actually begins to see decreasing returns as production levels decline, even though costs continue to increase.

In this example, the number of workers changed, while the land used, seeds planted, water consumed, and any other inputs remained the same. If more than one input were to change, the production results would vary and the law of diminishing returns may not apply if all of the inputs could be increased. If this case were to lead to increased production at lower average costs, economies of scale would be realized.

Table 1

Number of Workers	Corn Produced	Marginal Benefit
1	10	10
2	25	15
3	45	20
4	60	15
5	70	10
6	60	10

The concept of diminishing returns is as important for individuals and society as it is for businesses because it can have far-reaching effects on a wide variety of things, including the environment. This principle – although first thought to apply only to agriculture – is now widely accepted as an economic law that underlies all productive endeavors, including resource use and the cleanup of pollution.

The theory was effectively applied by Garrett Hardin in his 1968 article on the tragedy of the commons in which he looked at many common property resources, such as air, water, and forests, and described their use as being subject to diminishing returns. It is in this case that individuals acting in their own self-interest may "overuse" a resource because they do not take into consideration the impact it will have on a larger, societal scale. It can also be expanded to include limitations on our common resources. The services that fixed natural resources are able to provide – for example, in acting as natural filtration systems – will begin to diminish as contaminants and pollutants in the environment continue to increase. It is externalities such as these that can lead to the depletion of our resources and/or create other environmental problems.

However, the point at which diminishing returns can be illustrated is often very difficult to pinpoint because it varies with improved production techniques and other factors. In agriculture, for example, the debate about adequate supply remains unclear due to the uneven distribution of population and agricultural production around the globe and improvement in agricultural technology over time.

The challenge – whether it be local, regional, national, or global – is how best to manage the problem of declining resource-to-people ratios that could lead to a reduced standard of living. Widely used solutions for internalizing potential externalities include taxes, subsidies, and quotas. Often, there are attempts to find "bigger picture" solutions that focus on what many see as the primary causes, namely population growth and resource scarcity. Reducing population growth, along with increased technological innovation, may slow the growth in resource use and possibly offset the impact of diminishing returns. These potential benefits are a key reason why population growth and technological innovation are most often used in analyzing sustainable development possibilities.

Carrying Capacity

Changes in population can have a variety of economic, ecological, and social implications. One population issue is that of carrying capacity – the number of individuals an ecosystem can support without having any negative effects. It also includes a limit of resources and pollution levels that can be maintained without experiencing high levels of change. If carrying capacity is exceeded, living organisms must adapt to new levels of consumption or find alternative resources. Carrying capacity can be affected by the size of the human population, consumption of resources, and the level of pollution and environmental degradation that results. Carrying capacity, however, need not be fixed and can be expanded through good management and the development of

new resource-saving technologies. The relationship between carrying capacity and population growth has long been controversial. One of the original arguments appeared in 1798 by English economist Thomas Malthus who stated that continued population growth would cause over-consumption of resources. Malthus further argued that population was likely to grow at an exponential rate while food supplies would increase at an arithmetic rate, not keeping up with the exponential population growth. Malthus believed that an ever increasing population would continually strain society's ability to provide for itself and, as a result, mankind would be doomed to forever live in poverty.

Over a century later, American economist Julian Simon countered Malthus' arguments, asserting that an increase in population would improve the environment rather than degrade it. He believed human intellect to be the most valuable renewable natural resource that would continue to find innovative solutions to any problems that might arise – environmental, economical, or otherwise. Simon was also one of the founders of free market environmentalism, finding that a free market, together with appropriate property rights, was the best tool in order to preserve both the health and sustainability of the environment.

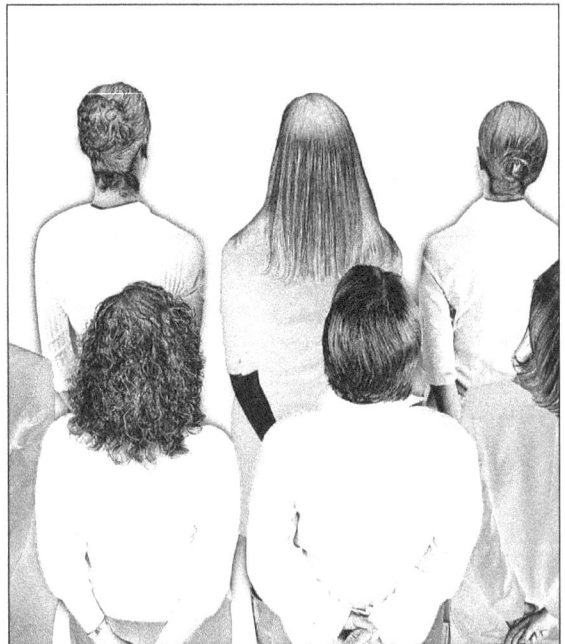

Figure 1

Throughout the late 1960s and 1970s, the controversy over the effect that an increasing population has on the Earth's on overpopulation, believed that human population had already exceeded the carrying capacity. Hardin is best known for his paper *The Tragedy of the Commons*, in which he argues that overpopulation of any species will deplete shared natural resources. Ehrlich, who wrote *The opulation Bomb* in 1968, predicted a population explosion accompanied by increasing famine and starvation. Although his prediction did not come true – in fact, in 1970 there was a slight decline in the population growth rate – he was correct in pointing out that, with the exception of solar energy, the Earth is a closed system with limited natural resources.

The standard of living in a region can help to alter an area's carrying capacity. Areas with a higher standard of living tend to have a reduced carrying capacity compared to areas with a lower standard of living due to the access to and demand for more resources. Nevertheless, the environmental Kuznets Curve – an observed phenomenon – suggests that beyond some point, increased income and environmental improvement often goes hand-in-hand. While population growth rates have stabilized and, in fact, are declining in many developed nations, consumption of resources and the generation of pollution and waste continue to grow. The effect this has on an ecosystem is called an "ecological footprint," which can be used to measure and manage the use of resources throughout an economy. It is also widely used as an indicator of environmental sustainability.

Carrying capacity often serves as the basis for sustainable development policies that attempt to balance the needs of today against the resources that will be needed in the future. The 1995 World Summit on Social Development defined sustainability as 'the framework to achieve a higher quality of life for all people in which economic development, social development, and environmental protection are interdependent and mutually beneficial components'. The 2002 World Summit furthered the process by identifying three key objectives of sustainable development: eradicating poverty, protecting natural resources, and changing unsustainable production and consumption patterns.

While the exact value of the human carrying capacity is uncertain and continues to be under debate, there has been evidence of the strain that both overpopulation and over-consumption has placed on some societies and the environment. Economists, ecologists, and policy analysts continue to study global consumption patterns to determine what the human carrying capacity is and what steps can be taken to ensure it is not exceeded. In the meantime, actions to reduce the strain and ensure natural resource recovery for the future will depend on an increase of sustainable development policies worldwide.

Two basic terms that are used most often by economists are *supply* and *demand*. How much of something that is available – the supply – and how much of something people want–the demand – are what makes a working market. Markets have existed since early in history when people bartered and made exchanges for food, trinkets, and other goods.

The market is the way in which an economic activity is organized between buyers and sellers through their behaviour and interaction with one another. Buyers, as a group, determine the overall demand for a particular product at various prices while sellers, as a group, determine the supply of a particular product at various prices.

The interaction of buyers and sellers in the market helps to determine the market price, thereby allocating scarce goods and services efficiently. The price is taken into account when deciding how much of something to consume, and also how much to produce. The relationship between price and quantity demanded is so universal that it is called the *law of demand*. This law states that with all else equal, when the price of a good rises, the quantity demanded falls – and when the price falls, the quantity demanded rises. The *law of supply* is just the opposite: the higher the price, the higher the quantity supplied – and the lower the price, less quantity is supplied.

A key function of the market is to find the equilibrium price when supply and demand are in balance. At this price, the goods supplied are equal to what is being demanded thereby bringing about the most efficient allocation of the goods. An efficient allocation of goods in a market is one in which no one can be made better off unless someone else is made worse off.

There are influences other than price, however, that often play a role in keeping the market from being truly efficient and at equilibrium. On the demand side, income can clearly play a significant role. As income rises, people will buy more of some goods or even begin to purchase higher quality – or more expensive – goods. The price of related goods can also alter demand. If the price of one cereal increases, for example, demand will likely switch to a similar cereal – which would be considered a substitute good. If the goods are considered to be complimentary – or are typically used together – a decrease in the price of one of the goods will increase the demand for another. An example of complimentary goods would be cars and gasoline where the price of gasoline depends partly on the number of cars. Personal tastes and expectations of the future also influence individual demands as does the number of buyers (an increase in buyers vying for a specific number of goods will increase the demand and likely increase the overall purchase price).

Table 2

Variables that Influence Buyers (Demand)	Variables that Influence Sellers (Supply)
Price	Price
Income	Input prices
Prices of related goods	Technology
Tastes	Expectations
Expectations	Number of sellers
Number of Buyers	

On the supply side, both expectations and the number of sellers can influence the number of goods produced. In addition, the cost of producing the good – or the input prices–as well as the level of technology used to turn the inputs into goods greatly influence the final price and quantity supplied.

Although most economic analyses focus on finding the market equilibrium, there exist a number of other market forms. When it comes to the utilization of natural resources or other environmental quality amenities, it is often difficult to find the equilibrium through mere market pricing since they are not true market goods. Efficiency would require maximizing current costs and benefits of using or extracting natural resources while also taking into consideration future costs and benefits, as well as the intrinsic and existence value of the resources. When the market fails to allocate the resources efficiently, market failure can occur. One example of this is the creation of externalities. Often, this occurs when clear property rights are absent, as with air and some water resources. Sometimes the government intervenes in an attempt to promote efficiency and bring the market back into equilibrium. Market options can include economic incentives and disincentives, or the establishment of property rights.

Net Present Value

Economists focus much of their analyses on a marketplace where supply and demand are based on the perceptions of present value and scarcity. However, when going beyond the simplicity of the short-term, particularly when costs and benefits occur at different points in time, it is important to utilize discounting to undertake longer-term analyses. Discounting adjusts costs and benefits to a common point in time. This approach can be useful in helping to determine how best to utilize many of our non-renewable natural resources.

Net present value (NPV) is a calculation used to estimate the value – or net benefit – over the lifetime of a particular project, often longer-term investments, such as building a new town hall or installing energy efficient appliances. NPV allows decision makers to compare various alternatives on a similar time scale by converting all options to current dollar figures. A project is deemed acceptable if the net present value is positive over the expected lifetime of the project.

The formula for NPV requires knowing the likely amount of time (t, usually in years) that cash will be invested in the project, the total length of time of the project (N, in the same unit of time as t), the interest rate (i), and the cash flow at that specific point in time (cash inflow – cash outflow, C).

$$NPV = \sum_{t=0}^{N} \frac{C_t}{(1+i)^t}$$

For example, take a business that is considering changing their lighting from traditional incandescent bulbs to fluorescents. The initial investment to change the lights themselves would be $40,000. After the initial investment, it is expected to cost $2,000 to operate the lighting system but will also yield $15,000 in savings each year; thus, there is a yearly cash flow of $13,000 every year after the initial investment. For simplicity, assume a discount rate of 10 per cent and an assumption that the lighting system will be utilized over a 5 year time period. This scenario would have the following NPV calculations:

$t = 0$ NPV $= (-40{,}000)/(1 + 0.10)\ 0 = -40{,}000.00$

$t = 1$ NPV $= (13{,}000)/(1.10)\ 1 = 11{,}818.18$

$t = 2$ NPV $= (13{,}000)/(1.10)\ 2 = 10{,}743.80$

$t = 3$ NPV $= (13{,}000)/(1.10)\ 3 = 9{,}767.09$

$t = 4$ NPV $= (13{,}000)/(1.10)\ 4 = 8{,}879.17$

$t = 5$ NPV $= (13{,}000)/(1.10)\ 5 = 8{,}071.98$

Based on the information above, the total net present value over the lifetime of the project would be $9,280.22.

Once the net present value is calculated, various alternatives can be compared and/or choices can be made. Any proposal with a NPV < 0 should be dismissed because it means that a project will likely lose money or not create enough benefit. The clear choice is a project whose NPV > 0 or, if there are several alternatives with positive NPVs, the choice would be the alternative with the higher NPV. With most societal choices, the opportunity costs are also considered when making decisions. Net present value provides one way to minimize foregone opportunities and identify the best possible options.

This particular example assumes that the interest rate does not change over time. Longer periods of time will often require separate calculations for each year in order to adjust for anticipated changes in the interest rate. When discounting is used it takes into account the fact that benefits in the future are not expected to be worth as much as in the present time. For example, $10 today may only be worth $9, $5, or even $1 in 2025. The rationale behind using a discount rate is two-fold: all things being equal, (1) individuals prefer to benefit now rather than later and (2) they tend to be risk averse, uncertain of what will occur in the future.

Net present value calculations can also help account for depreciation. Over time most assets depreciate, or lose value. Companies or individuals must be able to calculate a rate that includes depreciation for account balancing and tax purposes, as well to help predict replacement times for the asset in question. NPV and depreciation calculations are extremely valuable in the world of economics; they tell us what projects and businesses are better investments and what outcomes we may expect in the future.

However, while depreciation rates can be reliably estimated for most physical items, such as computer equipment or buildings, their application to natural resources and other environmental issues is more uncertain. Natural resources do not necessarily lose value over time. Thus, in most cases natural resources should not be depreciated when calculating resource NPVs. Also, since there is uncertainty about the future and external effects exist, it is much easier to predict what a company can do and what the reaction will be in the structured world of business than to accurately assess, say, the value of a forest to a local economy in future years.

Ecosystem Valuation

Valuation can be a useful tool that aids in evaluating different options that a natural resource manager might face. Because our ecological resources and services are so varied in their composition, it is often difficult to examine them on the same level. However, after they are assigned a value, an environmental resource or service can then be compared to any other item with a respective value. Ecosystem valuation is the process by which policymakers assign a value – monetary or otherwise – to environmental resources or to the outputs and/or services provided by those resources. For example, a mountain forest may provide environmental services by preventing downstream flooding.

Environmental resources and/or services are particularly hard to quantify due to their intangible benefits and multiple value options. It is almost impossible to attach a specific value to some of the experiences we have in nature, such as viewing a beautiful sunset. Problems also exist when a resource can be used for multiple purposes, such as a tree – the wood is valued differently if it is used for flood control versus if it is used for building a house. The quantity of a resource must also be taken into consideration because value can change depending how much of a resource is available. An example of this might be in preventing the first "unit" of pollution if we have a pristine air environment. Preventing the first unit of pollution is not valued very highly because the environment can easily recover. However, if the pollution continues until the air is becoming toxic to its surroundings; the

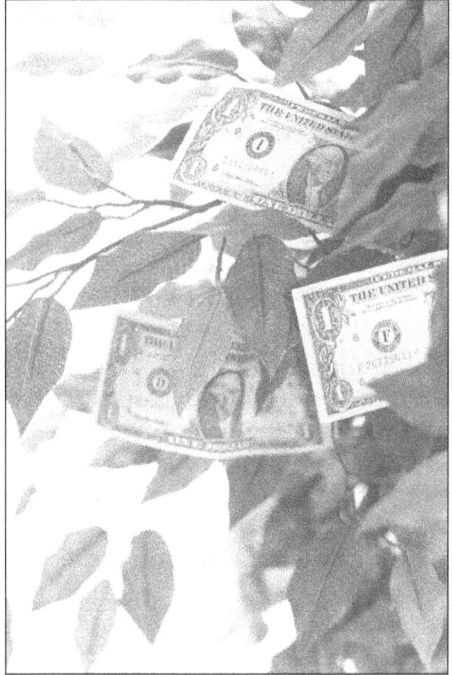

Figure 2

value of preserving clean air by preventing additional pollution is going to be increasingly valued.

Within economics, value is generally defined as the amount of alternate goods a person is willing to give up in order to get one "additional unit" of the good in question. An individual's preference for certain goods may either be stated or revealed. In the case of stated preferences, the amount of money a person is willing to pay for a good determines the value because that money could otherwise be used to purchase other goods. However, value may also be determined by simply ranking the alternatives according to the amount of benefit each will produce. Revealed preferences can be measured by examining a person's behaviour when it is not possible to use market pricing.

There are typically two ways to assign value to environmental resources and services – use and non-use – and there are approaches to measuring environmental benefits based on these defined values. When environmental resources or services are being used, it is easier to observe the price consumers are willing to pay for the conservation or preservation of those resources. Market or opportunity cost pricing can be used when there are tangible products to measure, such as the amount of fish caught in a lake. Replacement cost can also be used, calculated based on any expenses incurred to reverse environmental damage. Hedonic pricing will measure the effect that negative environmental qualities have on the price of related market goods. When evaluating non-use value, contingent valuation is employed through the use of surveys that attempt to assess an individual's willingness to pay for a resource that they do not consume.

A cost-benefit analysis requires the quantification of possible impacts of a proposed project. The impacts could be physical or monetary, but both must be calculated and included since a financial analysis that requires assigning dollar values to every resource evaluated is also performed. The process of environmental resource or service valuation provides a way to compare alternative proposals, but it is not without problems. All valuation techniques encompass a great deal of uncertainty: flaws can exist in the methods of assigning value accurately due to a wide number of variables and it is difficult to compartmentalize and measure environmental and natural resources and/or services within an ecosystem that functions as an interconnected web.

In summary, ecosystem valuation is a complex process by which economists attempt to assign a value to natural resources or to the ecological outputs and/or services provided by those resources. Although challenging, it allows policymakers to make decisions based on specific comparisons, typically monetary, rather than some other arbitrary basis. In recent years, the government has placed increasing emphasis on cost-effective laws and projects. Therefore, establishing a common measure by which to evaluate alternatives is essential.

As we make everyday choices – how much time to spend working or studying, what to spend our money on – we are experiencing what in economics are called trade-offs and opportunity costs. A trade-off is when we choose one option in favour of another and the opportunity cost is what is sacrificed in order to get something. Whether we realize it or not, we are constantly evaluating the costs and benefits of each decision we make; therefore, it can also be said that we are performing our own cost-benefit analysis each time we make a choice.

As decisions are made – either individually or as a society – we constantly make trade-offs in order to get more of one thing by giving up another. The saying "time is money" illustrates this point. If we 'consume' more free time, we are left with less money due to the fact that we are not earning money from using the time to work. The opposite is true as well; if we want more money, we must put in more work hours to get it; therefore there is less free time available. When we consider time and money, and graph the combinations for where one has no preference of one over the other, we come up with an indifference curve, such as the one below.

On the graph, X is the point where we have an even balance of time and money; yet an indifference curve is such that one is equally satisfied at any point along the curve. Therefore, we could move to point A, where we would have a lot more time but less money, or we could move to point B, with a lot more money but less time, and we would be equally satisfied. The slope of the indifference curve is based on the marginal utility of each decision; each successive move towards an axis comes at a higher price. For example, at point B we require more money for each unit of time than we do at point X because our time is more valuable since we have less of it. Therefore, we will begin to experience diminishing marginal utility.

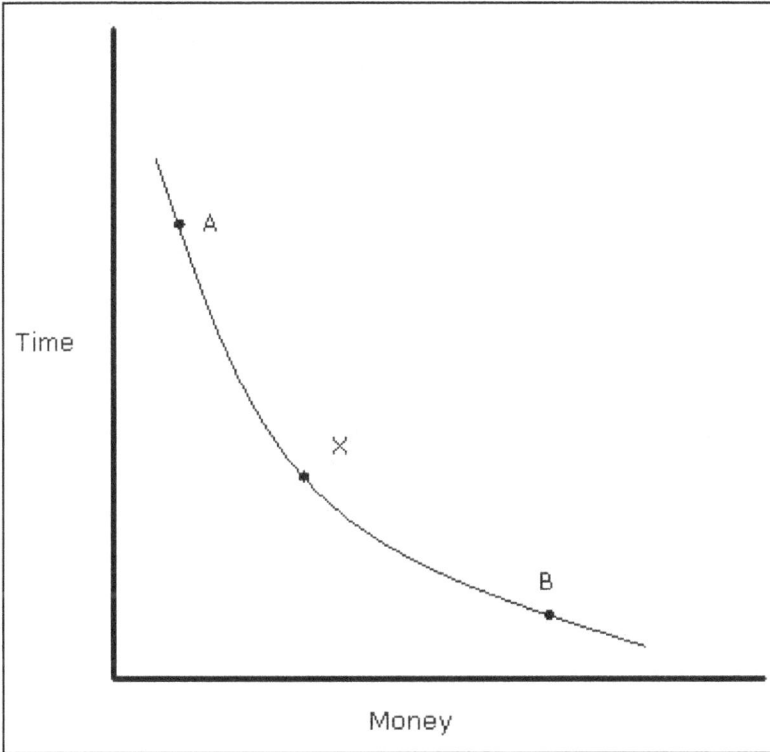

Figure 3

The economy and the environment are inextricably linked. Whether one is looking at daily life or natural resources and other environmental issues, because resources are scarce, choices have to be made about how to use them. The basic fact is that resources used to meet one choice or alternative cannot be used to meet another. Just like how we value regular goods, the valuation of natural resources and the environment is based on how we value their services and, for services that are consumed directly, that value is based on our utility and willingness to pay for a certain amount of the services.

The decision about how to allocate resources relating to the environment has an impact on all sectors of our economy, primarily because of the complex relationship between utilizing natural resources and economic output. Many times, the cost of utilizing these resources and/or services include direct costs as well as opportunity costs and external costs, which are not traded in markets or assessed directly in monetary terms. For example, when trees are cut for such uses as housing and furniture, some of the direct costs will include the cost of machinery and labour during cutting, processing, and manufacturing. The opportunity costs relating to this use would be the opportunities foregone by the machinery and labour that could not be used elsewhere, since it was occupied cutting trees. The external costs are the loss of environmental benefits that are no longer realized which may include a loss in watershed management services, species protection, and CO_2 reduction.

Many agree that in most cases the market is the best way to determine the allocation of resources. The demand for various products and the availability of natural resources – along with a number of other factors, including preferences, the number of buyers and sellers, pricing, alternative choices, etc. – is expected to lead to an efficient result of actual supply and demand. However, markets can fail to account for the full cost of a natural resource and/or services, which will prevent it from achieving an efficient allocation of the resource, leading to externalities.

To reduce the potential for market failures and their resulting externalities, planners and policymakers attempt to identify a course of action that generates the greatest societal benefits. Much of this is done by using a mix of policy and strategies, including regulation, taxes, permits, access restrictions, etc. It is finding the appropriate balance between utilizing our natural resources and meeting the demands of society that will allow us to continue to expand our economy while sustaining our natural resources and the environment.

Marginal costs and benefits are essential information for economists, businesses, and consumers. Even if we do not realize it, we all make decisions based on our marginal evaluations of the alternatives. In other words, "what does it cost to produce one more unit?" or "what will be the benefit of acquiring one more unit?"

When necessary, individual and social marginal cost and benefit curves can be drawn separately in order to understand different effects that a given action or policy might have. In the case of pollution, the social cost is generally higher than the individual cost due to externalities. However, as a whole, an economic system is considered efficient at the point where marginal benefit and marginal cost intersect, or are equal. Similar to the production of goods and services, we can utilize the same information in order to analyze pollution abatement – in terms of the *production* or *reduction* of pollution – within the market. In order to assess environmental improvement, we must take cost into consideration. The cost of these improvements is often thought of as the direct cost of any action taken in order to improve the environment.

Marginal cost measures the change in cost over the change in quantity. For example, if a company is producing 10 units at $100 total cost, and steps up production to 11 units at $120 total cost, the marginal cost is $20 since only the last unit of production is measured in order to calculate marginal cost. Mathematically speaking, it is the derivative of the total cost. Marginal cost is an important measurement because it accounts for increasing or decreasing costs of production, which allows a company to evaluate how much they actually pay to 'produce' one more unit.

Marginal cost will normally initially decrease through a short range, but increase as more is produced. Therefore the marginal cost curve is typically thought of as upward sloping. The marginal cost curve can represent a wide range of activities that can reduce the effects of environmental externalities, like pollution. The key point is that most environmental improvements are not free; resources must be expended in order for improvement to occur. For example, take an environment that has been polluted – while the initial *unit* of cleanup may be cheap, it becomes more and more expensive as additional cleanup is done. If cleanup is undertaken to point "Q", the total cost of the cleanup is P*Q the white and light gray areas on the graph below.

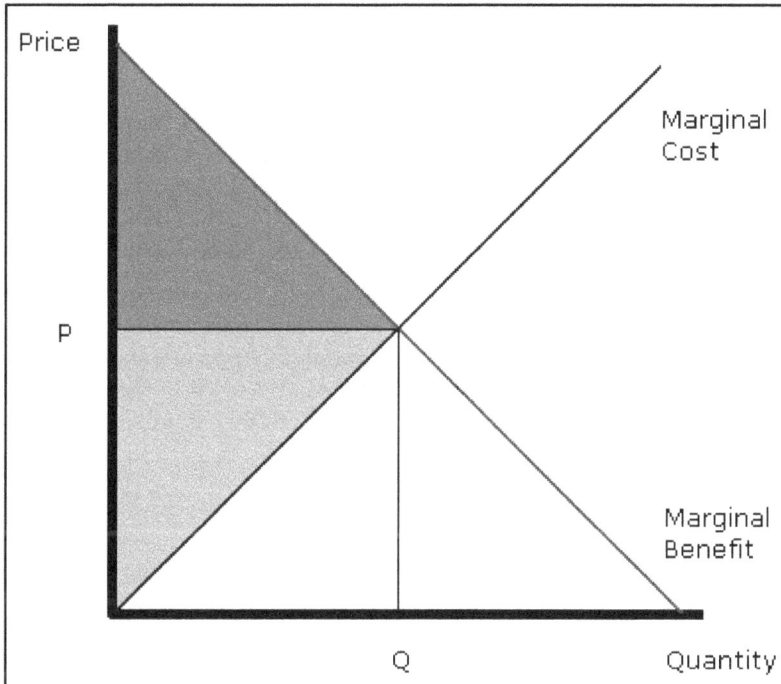

Figure 4

Marginal benefit is similar to marginal cost in that it is a measurement of the change in benefits over the change in quantity. While marginal cost is measured on the producer's end, marginal benefit is looked at from the consumer's perspective – in this sense it can be thought of as the demand curve for environmental improvement. The marginal benefit curve represents the tradeoff between environmental improvement and other things we could do with the resources needed to gain the improvement.

Again take an environment that has been polluted, the first unit of this pollution that is cleaned up has a very high benefit value to consumers of the environment. Each additional unit that is cleaned up is valued at a somewhat lower level than each previous one because the overall pollution level continues to decrease. Once the pollution is reduced below a certain point, the marginal benefit of additional pollution control measures will be negligible because the environment itself is able to absorb a low level of pollution. Taking a look at the graph above, the total consumer benefit that is represented as the dark grey area, the net benefit is greatest when the quantity – "Q" – reaches the marginal benefit curve. We could increase total benefit by adding pollution controls beyond Q, but only with marginal costs (MC) greater than marginal benefits (MB), so it is no longer efficient to further increase the benefits.

Oftentimes, benefits are more difficult to measure because they are not always monetary. In cases such as these the measurement may involve utilizing revealed preferences, through a survey or another mechanism, in order to discover the maximum price consumers are willing to pay for a particular quantity of a good. An *average* benefit is used when considering society as a whole because each individual's willingness to pay is different.

Marginal costs and benefits are a vital part of economics because they help to provide the relevant measurement of costs and benefits at a certain level of production and consumption. If measured marginal costs and benefits are provided, it is much easier to calculate the ideal price and quantity. It is where the two intersect that will always be the most economically efficient point of production and consumption.

When considering environmental issues, the efficient point at which marginal costs and marginal benefits are equal is an important economic concept because it captures the essence of tradeoffs. Often, environmental improvement concerns often revolve around whether we are above or below this point –whether any additional environmental improvement can provide more benefit than it will cost; this becomes an essential component in cost-benefit analysis.

Chapter 3

Externalities

Meaning, Sources and Types

Externalities arise when certain actions of producers or consumers have unintended external (indirect) effects on other producers or/and consumers. Externalities may be positive or negative. Positive externality arises when an action by an individual or a group confers benefits to others. A technological spillover is a positive externality and it occurs when a firm's invention not only benefits the firm but also enters into the society's pool of technological knowledge and benefits the society as a whole. Negative externalities arise when an action by an individual or group produces harmful effects on others. Pollution is a negative externality. When a factory discharges its untreated effluents in a river, the river is polluted and consumers of the river water bear costs in the form of health costs or/and water purification costs. In an activity generating positive externality, social benefit is higher than private benefit and in an activity generating negative externality, social cost is higher than private cost. Thus, in the presence of externalities, social benefits (costs) and private benefits (costs) differ.

The divergence between private benefits (costs) and social benefits (costs) results in inefficiency in resource allocation. Producers of externalities do not have any incentive to take into account the effects of their actions on others. In a competitive market economy, private optimum output is determined at the point where marginal private cost equals price. When a positive externality occurs, the marginal social benefit will be higher than the marginal private benefit (price) and hence the private optimal output will be lower than the social optimal output. When a negative externality occurs the marginal social cost will be higher than the marginal private cost (price) and hence the private optimal level of output will be higher than the social optimal output. Government intervention is needed to internalize externalities in production and consumption decisions of individuals so that social optimal levels of outputs and private optimal levels of outputs will be the same.

Environmental Externalities

The Environment (Protection) Act, 1986 defines environment to include 'water, air and land and the interrelationship which exists among and between water, air and land, and human beings, other living creatures, plants, microorganisms and property'.

Due to population growth and rapid industrialization, environmental resources such as groundwater and water in lakes and rivers and clean air in many places have become scarce resources. Industrial discharge of untreated effluents into water bodies and emissions into air have deteriorated the quality of water and air respectively. Negative intertemporal externalities occur when exhaustible resources are depleted and when renewable resources are harvested at rates greater than the regeneration rates.

Market Failures and Policy Failures

Environmental problems such as pollution and depletion and degradation of natural resources arise because of market failures and government failures. Market failures occur because markets for environmental goods and services do not exist or when the markets do exist, the market prices underestimate their social scarcity values.

Markets can exist and function efficiently only when property rights on goods and services exchanged are well defined and transaction costs of exchange are small.

For environmental resources such as clean air, water in river and springs, oceans and atmosphere, property rights are not well-defined. In most countries these resources are in public domain. Users of these resources consider them as "free" goods or "unpaid" factor of production. Therefore they impute zero prices for using these resources in their private decisions even when their social scarcity values are positive. Two important reasons for non-existence of the markets are (a) difficulty in defining, distributing and enforcing property rights and (b) high costs of creation and operation of markets.

Common property regimes do exist in fisheries and forestry. This regime permits exclusion of others from access to the resources. But collective ownership without binding agreements on rates of extraction on sharing of costs and benefits will result in overuse of the resources because each user does not take into account the consequences of his or her actions on other users. This is the well known problem of "tradegy of the commons".

Forestry provides both marketed (*e.g.*, timber) and non-marketed (*e.g.*, carbon sink) outputs and services. Forest owners/users will normally take into consideration in their private investment and production decisions only their revenues and costs and not the benefits which accrue to society as a whole, *e.g.*, biodiversity conservation, carbon sink and other ecosystem benefits. Here the markets are incomplete in the sense that there are no exchange institutions where the persons pay for the external benefits. As a result levels of activities such as habitat preservation, forest cover, biodiversity conservation will be below their social optimal levels.

Public ownership and management of common properties or regulation also pose problems because of lack of knowledge as well as information asymmetry both at the stage of design of rules and at the enforcement stage. Consideration of equity as well as political myopia act as barriers to social cost pricing of environmental resources and services.

Environmental Externalities: Types

Environmental externalities can be classified depending on how they affect individuals and regions. Environmental pollution or degradation may be local in nature as in water pollution in lakes, land degradation and air pollutant like particulate matter. Local pollution becomes a local public bad when it has two characteristics namely non-rivalry and non-exclusion. Pollution of large rivers and degradation of mountain ecosystems may affect many states/regions. Greenhouse gas emission is a global public bad in the sense that regardless of where the pollutants are emitted, the aggregate emissions affect all persons in the earth and the ecosystem as a whole.

The type of externality has a bearing on determining the appropriate unit for environmental governance. Here, the Subsidiarity Principle is relevant. This principle states that environmental decisions and enforcement be assigned to the lowest of government capable of handling it without significant residual externalities. For local public bads the appropriate authority is the concerned local body, that is, panchayat in village and municipality or corporation in an urban area. For pollution and natural degradation problems having regional effects the appropriate unit of governance may be state government. However, it should be noted that administrative boundaries and ecosystem boundaries may not match. For water resource development, an ecologically appropriate unit namely a watershed, may be within a state or it may be spread over more than one state. Environmental problems of transboundary nature, *e.g.*, acid rain, river pollution affecting more than one country and coastal zone degradation affect neighbouring countries. For global environmental problems such as climate change, ozone depletion and biodiversity loss, collective action level is needed at the global level.

Economics of Negative Production Externalities

Somewhere in the United States there is a steel plant located next to a river. This plant produces steel products, but it also produces "sludge," a by-product useless to the plant owners. To get rid of this unwanted by-product, the owners build a pipe out the back of the plant and dump the sludge into the river. The sludge produced is directly proportional to the production of steel; each additional unit of steel creates one more unit of sludge as well.

The steel plant is not the only producer using the river, however. Farther downstream is a traditional fishing area where fishermen catch fish for sale to local restaurants. Since the steel plant has begun dumping sludge into the river, the fishing has become much less profitable because there are many fewer fish left alive to catch.

This scenario is a classic example of what we mean by an externality. The steel plant is exerting a negative production externality on the fishermen, since its production adversely affects the well-being of the fishermen but the plant does not compensate the fishermen for their loss.

One way to see this externality is to graph the market for the steel produced by this plant (Figure 1) and to compare the private benefits and costs of production to the social benefits and costs. *Private benefits and costs* are the benefits and costs borne directly by the actors in the steel market (the producers and consumers of the steel products). *Social benefits and costs* are the private benefits and costs *plus* the benefits and costs to any actors outside this steel market who are affected by the steel plant's production process (the fishermen).

Each point on the market supply curve for a good (steel, in our example) represents the market's marginal cost of producing that unit of the good – that is, the private marginal cost (*PMC*) of that unit of steel. What determines the welfare consequences of production, however, is the social marginal cost (*SMC*), which equals the private marginal cost to the producers of producing that next unit of a good *plus any costs associated with the production of that good that are imposed on others*. This distinction was not made in Chapter 2, because without market failures $SMC = PMC$, the social costs of producing steel are equal to the costs to steel producers.

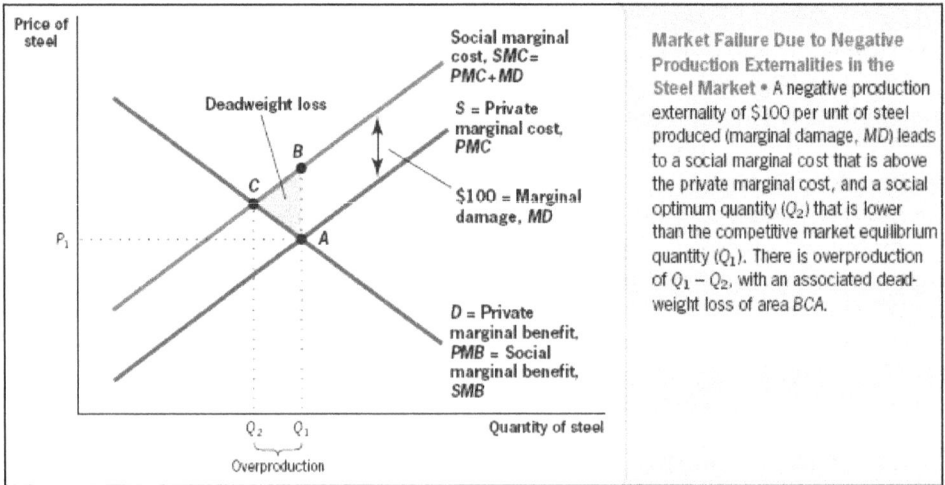

Figure 1

This approach is not correct in the presence of externalities, however. When there are externalities, $SMC = PMC + MD$, where MD is the marginal damage done to others, such as the fishermen, from each unit of production (marginal because it is the damage associated with that particular unit of production, not total production). Suppose, for example, that each unit of steel production creates sludge that kills $100 worth of fish. In Figure 2, the SMC curve is therefore the PMC (supply) curve, shifted upward by the marginal damage of $100.5 That is, at Q_1 units of production (point A), the social marginal cost is the private marginal cost at that point (which is equal to P_1), plus \$100 (point B). For every level of production, social costs are \$100 higher than private costs, since each unit of production imposes \$100 of costs on the fishermen for which they are not compensated.

Negative Consumption Externalities

It is important to note that externalities do not arise solely from the production side of a market. Consider the case of cigarette smoke. In a restaurant that allows smoking, your consumption of cigarettes may have a negative effect on my enjoyment of a restaurant meal. Yet you do not in any way pay for this negative effect on me. This is an example of a negative consumption externality, whereby consumption of a good reduces the well-being of others, a loss for which they are not compensated. When there is a negative consumption externality, $SMB = PMB - MD$, where MD is the marginal damage done to others by your consumption of that unit. For example, if MD is 40′ a pack, the marginal damage done to others by your smoking is 40′ for every pack you smoke.

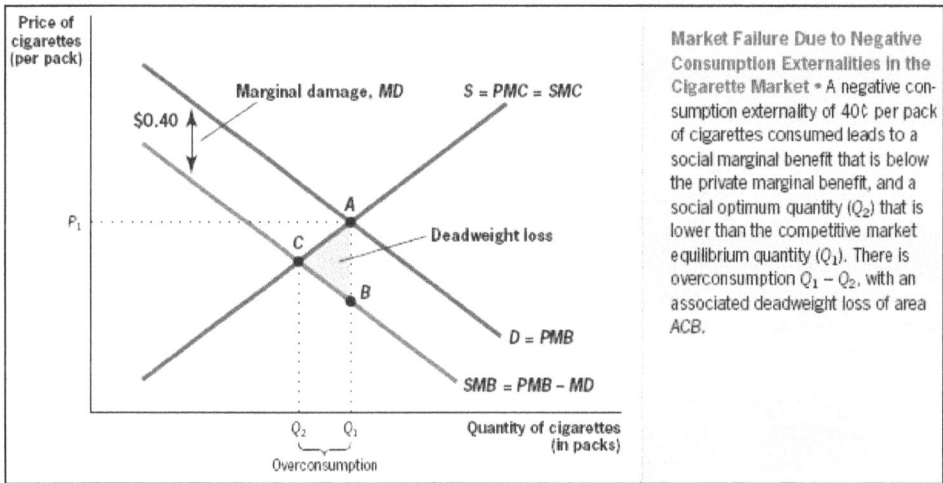

Figure 2

Figure 2 shows supply and demand in the market for cigarettes. The supply and demand curves represent the PMC and PMB. The private equilibrium is at point A, where supply (PMC) equals demand (PMB), with cigarette consumption of Q_1 and price of P_1. The SMC equals the PMC because there are no externalities associated with the production of cigarettes in this example. Note, however, that the SMB is now below the PMB by 40′ per pack; every pack consumed has a social benefit that is 40′ below its private benefit. That is, at Q_1 units of production (point A), the social marginal benefit is the private marginal benefit at that point (which is equal to P_1), minus 40′ (point B). For each pack of cigarettes, social benefits are 40′ lower than private benefits, since each pack consumed imposes 40′ of costs on others for which they are not compensated.

The social-welfare-maximizing level of consumption, Q_2, is identified by point C, the point at which $SMB = SMC$. There is overconsumption of cigarettes by $Q_2 - Q_2$: the social costs (point A on the SMC curve) exceed social benefits (on the SMB curve) for all units between Q_2 and Q_2. As a result, there is a deadweight loss (area ACB) in the market for cigarettes.

The Externality of SUVs

In 1985, the typical driver sat behind the wheel of a car that weighed about 3,200 pounds, and the largest cars on the road weighed 4,600 pounds. Today, the typical driver is in a car that weighs 4,089 pounds (an increase of 28 per cent) and the largest cars on the road can weigh 8,500 pounds. The major culprits in this evolution of car size are sport utility vehicles (SUVs). The term *SUV* was originally reserved for large vehicles intended for off-road driving, but it now refers to any large passenger vehicle marketed as an SUV, even if it lacks off-road capabilities. SUVs, with an average weight of 4,500 pounds, represented only 6.4 per cent of vehicle sales as recently as 1988, but 17 years later, in 2005, they accounted for over 25 per cent of the new vehicles sold each year.

The consumption of large cars such as SUVs produces three types of negative externalities:

Environmental Externalities

The contribution of driving to global warming is directly proportional to the amount of fossil fuel a vehicle requires to travel a mile. The typical compact or mid-size car gets roughly 25 miles to the gallon but the typical SUV gets only 18 miles to the gallon. This means that SUV drivers use more gas to go to work or run their errands, increasing fossil fuel emissions. This increased environmental cost is not paid by those who drive SUVs.

Wear and Tear on Roads Each year, federal, state, and local governments in the United States spend $33.2 billion repairing our roadways. Damage to roadways comes from many sources, but a major culprit is the passenger vehicle, and the damage it does to the roads is proportional to vehicle weight. When individuals drive SUVs, they increase the cost to government of repairing the roads. SUV drivers bear some of these costs through gasoline taxes (which fund highway repair), since the SUV uses more gas, but it is unclear if these extra taxes are enough to compensate for the extra damage done to roads.

Safety Externalities One major appeal of SUVs is that they provide a feeling of security because they are so much larger than other cars on the road. Offsetting this feeling of security is the added *insecurity* imposed on other cars on the road. For a car of average weight, the odds of having a fatal accident rise by four times if the accident is with a typical SUV and not with a car of the same size. Thus, SUV drivers impose a negative externality on other drivers because they don't compensate those other drivers for the increased risk of a dangerous accident.

Positive Externalities

When economists think about externalities, they tend to focus on negative externalities, but not all externalities are bad. There may also be positive production externalities associated with a market, whereby production benefits parties other than the producer and yet the producer is not compensated. Imagine the following scenario: There is public land beneath which there *might* be valuable oil reserves. The government allows any oil developer to drill in those public lands, as long as the government gets some royalties on any oil reserves found. Each dollar the oil developer spends on exploration increases the chances of finding oil reserves. Once found, however, the oil reserves can be tapped by other companies; the initial driller only has the advantage of getting there first. Thus, exploration for oil by one company exerts a *positive production externality* on other companies: each dollar spent on exploration by the first company raises the chance that other companies will have a chance to make money from new oil found on this land.

Figure 3 shows the market for oil exploration to illustrate the positive externality to exploration: the social marginal cost of exploration is actually *lower* than the private marginal cost because exploration has a positive effect on the future profits of other companies. Assume that the marginal benefit of each dollar of exploration by one company, in terms of raising the expected profits of other companies who drill the same land, is a constant amount MB. As a result, the SMC is below the PMC by the amount MB. Thus, the private equilibrium in the exploration market (point A, quantity Q_1) leads to *underproduction* relative to the optimal level (point B, quantity Q_2) because the initial oil company is not compensated for the benefits it confers on other oil producers.

Note also that there can be positive consumption externalities. Imagine, for example, that my neighbour is considering improving the landscaping around his house. The improved landscaping will cost him $1,000, but it is only worth $800 to

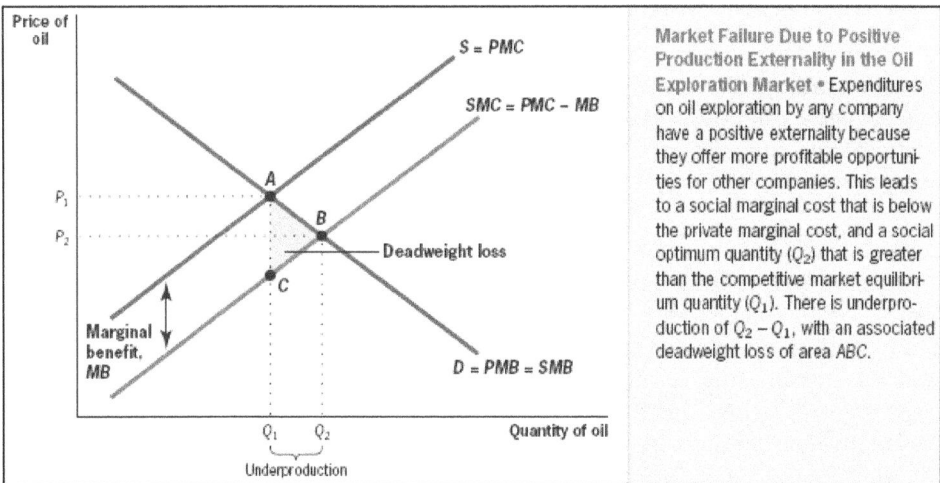

Market Failure Due to Positive Production Externality in the Oil Exploration Market • Expenditures on oil exploration by any company have a positive externality because they offer more profitable opportunities for other companies. This leads to a social marginal cost that is below the private marginal cost, and a social optimum quantity (Q_2) that is greater than the competitive market equilibrium quantity (Q_1). There is underproduction of $Q_2 - Q_1$, with an associated deadweight loss of area ABC.

Figure 3

him. My bedroom faces his house, and I would like to have nicer landscaping to look at. This better view would be worth $300 to me. That is, the total social marginal benefit of the improved landscaping is $1,100, even though the private marginal benefit to my neighbor is only $800. Since this social marginal benefit ($1,100) is larger than the social marginal costs ($1,000), it would be socially efficient for my neighbor to do the landscaping. My neighbour won't do the landscaping, however, since his private costs ($1,000) exceed his private benefits. His landscaping improvements would have a positive effect on me for which he will not be compensated, thus leading to an under consumption of landscaping.

Quick Hint One confusing aspect of the graphical analysis of externalities is knowing which curve to shift, and in which direction. To review, there are four possibilities:

☆ Negative production externality: *SMC* curve lies above *PMC* curve

☆ Positive production externality: *SMC* curve lies below *PMC* curve

☆ Negative consumption externality: *SMB* curve lies below *PMB* curve

☆ Positive consumption externality: *SMB* curve lies above *PMB* curve

Armed with these facts, the key is to assess which category a particular example fits into. This assessment is done in two steps. First, you must assess whether the externality is associated with producing a good or with consuming a good.

Then, you must assess whether the externality is positive or negative. The steel plant example is a negative production externality because the externality is associated with the production of steel, not its consumption; the sludge doesn't come from using steel, but rather from making it. Likewise, our cigarette example is a negative consumption externality because the externality is associated with the consumption of cigarettes; secondhand smoke doesn't come from making cigarettes, it comes from smoking them.

Private-Sector Solutions to Negative Externalities

In microeconomics, the market is innocent until proven guilty (and, similarly, the government is often guilty until proven innocent!). An excellent application of this principle can be found in a classic work by Ronald Coase, a professor at the Law School at the University of Chicago, who asked in 1960: Why won't the market simply compensate the affected parties for externalities?

The Solution

To see how a market might compensate those affected by the externality, let's look at what would happen if the fishermen owned the river in the steel plant example. They would march up to the steel plant and demand an end to the sludge dumping that was hurting their livelihood. They would have the right to do so because they have *property rights* over the river; their ownership confers to them the ability to control the use of the river.

Suppose for the moment that when this conversation takes place there is no pollution-control technology to reduce the sludge damage; the only way to reduce sludge is to reduce production. So ending sludge dumping would mean shutting down the steel plant. In this case, the steel plant owner might propose a compromise: she would pay the fishermen $100 for each unit of steel produced, so that they were fully compensated for the damage to their fishing grounds. As long as the steel plant can make a profit with this extra $100 payment per unit, then this is a better deal for the plant than shutting down, and the fishermen are fully compensated for the damage done to them.

This type of resolution is called internalizing the externality. Because the fishermen now have property rights to the river, they have used the market to obtain compensation from the steel plant for its pollution. The fishermen have implicitly created a market for pollution by pricing the bad behaviour of the steel plant. From the steel plant's perspective, the damage to the fish becomes just another input cost, since it has to be paid in order to produce.

This point is illustrated in Figure 4. Initially, the steel market is in equilibrium at point A, with quantity Q_1 and price P_1, where $PMB = PMC_1$. The socially optimal level of steel production is at point B, with quantity Q_2 and price P_2, where $SMB = SMC = PMC_1 + MD$. Because the marginal cost of producing each unit of steel has increased by $100 (the payment to the fishermen), the private marginal cost curve shifts upward from PMC_1 to PMC_2, which equals SMC. That is, social marginal costs are private marginal costs plus $100, so by adding $100 to the private marginal costs, we raise the PMC to equal the SMC. There is no longer overproduction because the social marginal costs and benefits of each unit of production are equalized. This example illustrates Part I of the Coase Theorem: when there are well-defined property rights and costless bargaining, then negotiations between the party creating the externality and the party affected by the externality can bring about the socially optimal market quantity. This theorem states that externalities do not necessarily create market failures, because negotiations between the parties can lead the offending producers (or consumers) to *internalize the externality*, or account for the external effects in their production (or consumption).

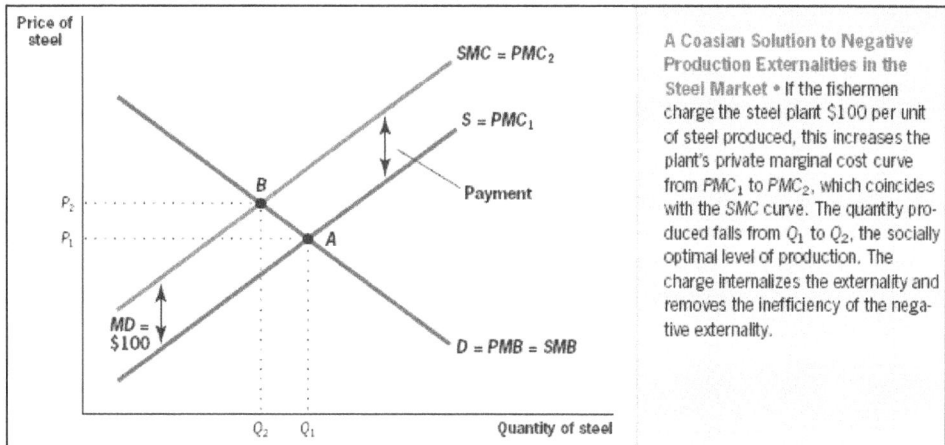

A Coasian Solution to Negative Production Externalities in the Steel Market • If the fishermen charge the steel plant $100 per unit of steel produced, this increases the plant's private marginal cost curve from PMC_1 to PMC_2, which coincides with the SMC curve. The quantity produced falls from Q_1 to Q_2, the socially optimal level of production. The charge internalizes the externality and removes the inefficiency of the negative externality.

Figure 4

The Coase theorem suggests a very particular and limited role for the government in dealing with externalities: establishing property rights. In Coase's view, the fundamental limitation to implementing private-sector solutions to externalities is poorly established property rights. If the government can establish and enforce those property rights, then the private market will do the rest.

The Coase theorem also has an important **Part II:** the efficient solution to an externality does not depend on which party is assigned the property rights, as long as someone is assigned those rights. We can illustrate the intuition behind Part II using the steel plant example. Suppose that the steel plant, rather than the fishermen, owned the river. In this case, the fishermen would have no right to make the plant owner pay a $100 compensation fee for each unit of steel produced. The fishermen, however,would find it in their interest to pay the steel plant to produce less. If the fishermen promised the steel plant owner a payment of $100 for each unit he did not produce, then the steel plant owner would rationally consider there to be an extra $100 cost to each unit he did produce. Remember that in economics, opportunity costs are included in a firm's calculation of costs; thus, forgoing a payment from the fishermen of $100 for each unit of steel not produced has the same effect on production decisions as being forced to pay $100 extra for each unit of steel produced. Once again, the private marginal cost curve would incorporate this extra (opportunity) cost and shift out to the social marginal cost curve, and there would no longer be overproduction of steel.

Quick Hint You may wonder why the fishermen would ever engage in either of these transactions: they receive $100 for each $100 of damage to fish, or pay $100 for each $100 reduction in damage to fish. So what is in it for them? The answer is that this is a convenient shorthand economics modelers use for saying, "The fishermen would charge at least $100 for sludge dumping" or "The fishermen would pay up to $100 to remove sludge dumping." By assuming that the payments are exactly $100, we can conveniently model private and social marginal costs as equal. It may be useful for you to think of the payment to the fishermen as $101 and the payment from the fishermen as $99, so that the fishermen make some money and private and social costs are approximately equal. In reality, the payments to or from the fishermen will depend on the negotiating power and skill of both parties in this transaction, highlighting the importance of the issues raised next.

The Problems with Coasian Solutions

This elegant theory would appear to rescue the standard competitive model from this important cause of market failures and make government intervention unnecessary (other than to ensure property rights). In practice, however, the Coase theorem is unlikely to solve many of the types of externalities that cause market failures.We can see this by considering realistically the problems involved in achieving a "Coasian solution" to the problem of river pollution.

The Assignment Problem

The first problem involves assigning blame. Rivers can be very long, and there may be other pollution sources along the way that are doing some of the damage to the fish. The fish may also be dwindling for natural reasons, such as disease or a rise in natural predators. In many cases, it is impossible to assign blame for externalities to one specific entity.

Assigning damage is another side to the assignment problem. We have assumed that the damage was a fixed dollar amount, $100. Where does this figure come from in practice? Can we trust the fishermen to tell us the right amount of damage that they suffer? It would be in their interest in any Coasian negotiation to overstate the damage in order to ensure the largest possible payment. And how will the payment be distributed among the fishermen? When a number of individuals are fishing the same area, it is difficult to say whose catch is most affected by the reduction in the stock of available fish.

The significance of the assignment problem as a barrier to internalizing the externality depends on the nature of the externality. If my loud stereo playing disturbs your studying, then assignment of blame and damages is clear. In the case of global warming, however, how can we assign blame clearly when carbon emissions from any source in the world contribute to this problem? And how can we assign damages clearly when some individuals would like the world to be hotter, while others would not? Because of assignment problems, Coasian solutions are likely to be more effective for small, localized externalities than for larger, more global externalities.

The Holdout Problem

Imagine that we have surmounted the assignment problem and that by careful scientific analysis we have determined that each unit of sludge from the steel plant kills $1 worth of fish for each of 100 fishermen, for a total damage of $100 per unit of steel produced.

Now, suppose that the fishermen have property rights to the river, and the steel plant can't produce unless all 100 fishermen say it can. The Coasian solution is that each of the 100 fishermen gets paid $1 per unit of steel production, and the plant continues to produce steel. Each fisherman walks up to the plant and collects his check for $1 per unit. As the last fisherman is walking up, he realizes that he suddenly has been imbued with incredible power: the steel plant cannot produce without his permission since he is a part owner of the river. So, why should he settle for only $1 per unit? Having already paid out $99 per unit, the steel plant would probably be willing to pay more than $1 per unit to remove this last obstacle to their production. Why not ask for $2 per unit? Or even more?

This is an illustration of the holdout problem, which can arise when the property rights in question are held by more than one party: the shared property rights give each party power over all others. If the other fishermen are thinking ahead they will realize this might be a problem, and they will all try to be the last one to go to the plant. The result could very well be a breakdown of the negotiations and an inability to negotiate a Coasian solution. As with the assignment problem, the holdout problem would be amplified with a huge externality like global warming, where billions of persons are potentially damaged.

The Free Rider Problem

Can we solve the holdout problem by simply assigning the property rights to the side with only one negotiator, in this case the steel plant? Unfortunately, doing so creates a new problem.

Suppose that the steel plant has property rights to the river, and it agrees to reduce production by 1 unit for each $100 received from fishermen. Then the Coasian solution would be for the fishermen to pay $100, and for the plant to then move to the optimal level of production. Suppose that the optimal reduction in steel production (where social marginal benefits and costs are equal) is 100 units, so that each fisherman pays $100 for a total of $10,000, and the plant reduces production by 100 units.

Suppose, once again, that you are the last fisherman to pay. The plant has already received $9,900 to reduce its production, and will reduce its production as a result by 99 units. The 99 units will benefit all fishermen equally since they all share the river. Thus, as a result, if you don't pay your $100, you will still be almost as well off in terms of fishing as if you do. That is, the damage avoided by that last unit of reduction will be shared equally among all 100 fishermen who use the river, yet you will pay the full $100 to buy that last unit of reduction. Thought of that way, why would you pay? This is an example of the free rider problem: when an investment has a personal cost but a common benefit, individuals will under invest. Understanding this incentive, your fellow fishermen will also not pay their $100, and the externality will remain unsolved; if the other fishermen realize that someone is going to grab a free ride, they have little incentive to pay in the first place.

Transaction Costs and Negotiating Problems

Finally, the Coasian approach ignores the fundamental problem that it is hard to negotiate when there are large numbers of individuals on one or both sides of the negotiation. How can the 100 fishermen effectively get together and figure out how much to charge or pay the steel plant? This problem is amplified for an externality such as global warming, where the potentially divergent interests of billions of parties on one side must be somehow aggregated for a negotiation.

Moreover, these problems can be significant even for the small-scale, localized externalities for which Coase's theory seems best designed. In theory, my neighbour and I can work out an appropriate compensation for my loud music disturbing his studying. In practice, this may be a socially awkward conversation that is more likely to result in tension than in a financial payment. Similarly, if the person next to me in the restaurant is smoking, it would be far outside the norm, and probably considered insulting, to lean over and offer him $5 to stop smoking. Alas, the world does not always operate in the rational way economists wish it would!

Bottom Line Ronald Coase's insight that externalities can sometimes be internalized was a brilliant one. It provides the competitive market model with a defense against the onslaught of market failures that we will bring to bear on it throughout this course. It is also an excellent reason to suspect that the market may be able to internalize some small-scale, localized externalities. Where it won't help, as we've seen, is with large-scale, global externalities that are the focus of, for example, environmental policy in the United States. The government may therefore have a role to play in addressing larger externalities.

Public-Sector Remedies for Externalities

In the United States, public policy makers do not think that Coasian solutions are sufficient to deal with large-scale externalities. The Environmental Protection Agency (EPA) was formed in 1970 to provide public-sector solutions to the problems of externalities in the environment. The agency regulates a wide variety of environmental issues, in areas ranging from clean air to clean water to land management.

Public policy makers employ three types of remedies to resolve the problems associated with negative externalities.

Corrective Taxation

We have seen that the Coasian goal of "internalizing the externality" may be difficult to achieve in practice in the private market. The government can achieve this same outcome in a straightforward way, however, by taxing the steel producer an amount MD for each unit of steel produced.

Figure 5 illustrates the impact of such a tax. The steel market is initially in equilibrium at point A, where supply ($=PMC_1$) equals demand ($= PMB = SMB$), and Q_1 units of steel are produced at price P_1. Given the externality with a cost of MD, the socially optimal production is at point B, where social marginal costs and benefits are equal. Suppose that the government levies a tax per unit of steel produced at an amount $t = MD$. This tax would act as another input cost for the steel producer, and would shift its private marginal cost up by MD for each unit produced. This will result in a new PMC curve, PMC_2, which is identical to the SMC curve. As a result, the tax effectively internalizes the externality and leads to the socially optimal outcome (point B, quantity Q_2). The government per unit tax on steel production acts in the same way as if the fishermen owned the river. This type of corrective taxation is often

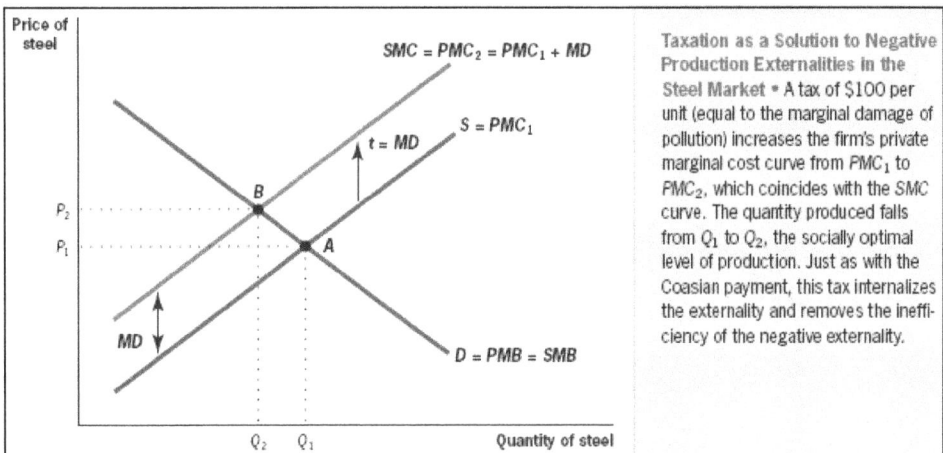

Taxation as a Solution to Negative Production Externalities in the Steel Market • A tax of $100 per unit (equal to the marginal damage of pollution) increases the firm's private marginal cost curve from PMC_1 to PMC_2, which coincides with the SMC curve. The quantity produced falls from Q_1 to Q_2, the socially optimal level of production. Just as with the Coasian payment, this tax internalizes the externality and removes the inefficiency of the negative externality.

Figure 5

called "Pigouvian taxation," after the economist A.C. Pigou, who first suggested this approach to solving externalities.

Subsidies

As noted earlier, not all externalities are negative; in cases such as oil exploration or nice landscaping by your neighbors, externalities can be positive.

The Coasian solution to cases such as the oil exploration case would be for the other oil producers to take up a collection to pay the initial driller to search for more oil reserves (thus giving them the chance to make more money from any oil that is found). But, as we discussed, this may not be feasible.

The government can achieve the same outcome by making a payment, or a subsidy, to the initial driller to search for more oil. The amount of this subsidy would exactly equal the benefit to the other oil companies and would cause the initial driller to search for more oil, since his cost per barrel has been lowered.

The impact of such a subsidy is illustrated in Figure 6, which shows once again the market for oil exploration. The market is initially in equilibrium at point A where PMC_1 equals PMB, and Q_1 barrels of oil are produced at price P_1. Given the positive externality with a benefit of MB, the socially optimal production is at point B, where social marginal costs and benefits are equal. Suppose that the government pays a subsidy per barrel of oil produced of $S = MB$. The subsidy would lower the private marginal cost of oil production, shifting the private marginal cost curve down by MB for each unit produced. This will result in a new PMC curve, PMC_2, which is identical to the SMC curve. The subsidy has caused the initial driller to internalize the positive externality, and the market moves from a situation of underproduction to one of optimal production.

Regulation

Throughout this discussion, you may have been asking yourself: Why this fascination with prices, taxes, and subsidies? If the government knows where the

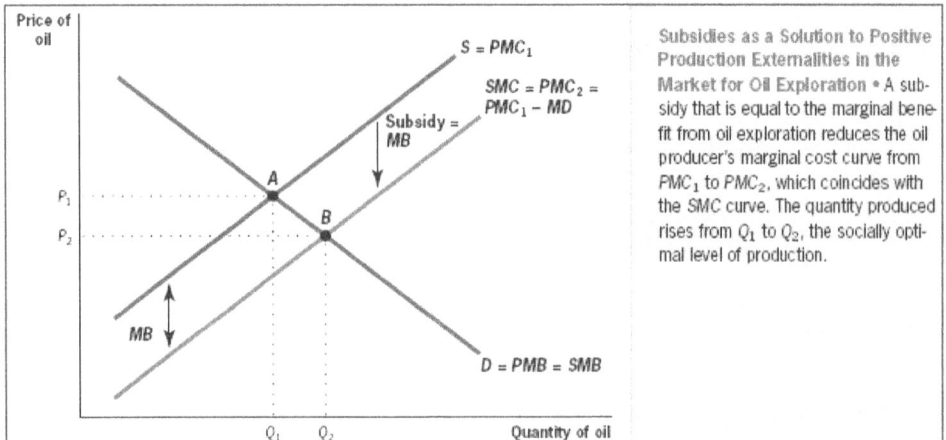

Subsidies as a Solution to Positive Production Externalities in the Market for Oil Exploration • A subsidy that is equal to the marginal benefit from oil exploration reduces the oil producer's marginal cost curve from PMC_1 to PMC_2, which coincides with the SMC curve. The quantity produced rises from Q_1 to Q_2, the socially optimal level of production.

Figure 6

socially optimal level of production is, why doesn't it just mandate that production take place at that level, and forget about trying to give private actors incentives to produce at the optimal point? Using Figure 6 as an example, why not just mandate a level of steel production of Q_2 and be done with it?

In an ideal world, Pigouvian taxation and regulation would be identical. Because regulation appears much more straightforward, however, it has been the traditional choice for addressing environmental externalities in the United States and around the world. When the U.S. government wanted to reduce emissions of sulphur dioxide (SO2) in the 1970s, for example, it did so by putting a limit or cap on the amount of sulfur dioxide that producers could emit, not by a tax on emissions. In 1987, when the nations of the world wanted to phase out the use of chlorofluorocarbons (CFCs), which were damaging the ozone layer, they banned the use of CFCs rather than impose a large tax on products that used CFCs.

Given this governmental preference for quantity regulation, why are economists so keen on taxes and subsidies? In practice, there are complications that may make taxes a more effective means of addressing externalities. In the next section, we discuss two of the most important complications. In doing so, we illustrate the reasons that policy makers might prefer regulation, or the "quantity approach" in some situations, and taxation, or the "price approach" in others.

Distinctions between Price and Quantity Approaches to Addressing Externalities

In this section, we compare price (taxation) and quantity (regulation) approaches to addressing externalities, using more complicated models in which the social efficiency implications of intervention might differ between the two approaches. The goal in comparing these approaches is to find the most efficient path to environmental targets. That is, for any reduction in pollution, the goal is to find the lowest-cost means of achieving that reduction.

Basic Model

To illustrate the important differences between the price and quantity approaches, we have to add one additional complication to the basic competitive market that we have worked with thus far. In that model, the only way to reduce pollution was to cut back on production. In reality, there are many other technologies available for reducing pollution besides simply scaling back production. For example, to reduce sulphur dioxide emissions from coal-fired power plants, utilities can install smokestack scrubbers that remove SO_2 from the emissions and sequester it, often in the form of liquid or solid sludge that can be disposed of safely. Passenger cars can also be made less polluting by installing "catalytic converters," which turn dangerous nitrogen oxide into compounds that are not harmful to public health.

To understand the differences between price and quantity approaches to pollution reduction, it is useful to shift our focus from the market for a good (*e.g.*, steel) to the "market" for pollution reduction, as illustrated in Figure 7. In this diagram, the horizontal axis measures the extent of pollution reduction undertaken by a plant; a value of zero indicates that the plant is not engaging in any pollution reduction. Thus, the horizontal axis also measures the amount of pollution: as you move to the right, there is more pollution reduction and less pollution. We show this by denoting *more reduction* as you move to the right on the horizontal axis; *Rfull* indicates that pollution has been reduced to zero. *More pollution* is indicated as you move to the left on the horizontal axis; at *Pfull*, the maximum amount of pollution is being produced. The vertical axis represents the cost of pollution reduction to the plant, or the benefit of pollution reduction to society (that is, the benefit to other producers and consumers who are not compensated for the negative externality).

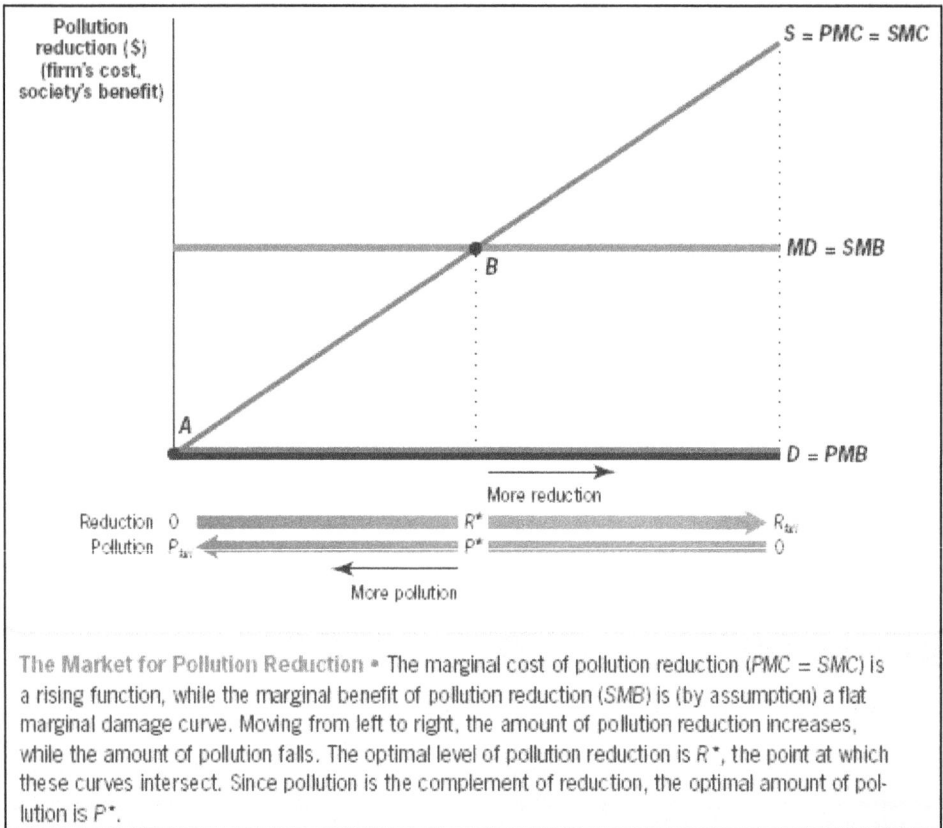

The Market for Pollution Reduction • The marginal cost of pollution reduction ($PMC = SMC$) is a rising function, while the marginal benefit of pollution reduction (SMB) is (by assumption) a flat marginal damage curve. Moving from left to right, the amount of pollution reduction increases, while the amount of pollution falls. The optimal level of pollution reduction is R^*, the point at which these curves intersect. Since pollution is the complement of reduction, the optimal amount of pollution is P^*.

Figure 7

The *MD* curve represents the marginal damage that is averted by additional pollution reduction. This measures the social marginal benefit of pollution reduction. Marginal damage is drawn flat at $100 for simplicity, but it could be downward sloping due to diminishing returns. The private marginal benefit of pollution reduction is zero, so it is represented by the horizontal axis; there is no gain to the plant's private interests from reducing dumping.

The *PMC* curve represents the plant's private marginal cost of reducing pollution. The *PMC* curve slopes upward because of diminishing marginal productivity of this input. The first units of pollution are cheap to reduce: just tighten a few screws or put a cheap filter on the sludge pipe. Additional units of reduction become more expensive, until it is incredibly expensive to have a completely pollution-free production process. Because there are no externalities from the production of pollution reduction (the externalities come from the end product, reduced pollution, as reflected in the *SMB* curve, not from the process involved in actually reducing the pollution), the *PMC* is also the *SMC* of pollution reduction.

The free market outcome in any market would be zero pollution reduction. Since the cost of pollution is not borne by the plant, it has no incentive to reduce pollution. The plant will choose zero reduction and a full amount of pollution *Pfull* (point *A*, at which the *PMC* of zero equals the *PMB* of zero).

What is the optimal level of pollution reduction? The optimum is *always* found at the point at which social marginal benefits and costs are equal, here point *B*. The optimal quantity of pollution reduction is R^*: at that quantity, the marginal benefits of reduction (the damage done by pollution) and the marginal costs of reduction are equal. Note that setting the optimal amount of pollution reduction is the same as setting the optimal amount of pollution. If the free market outcome is pollution reduction of zero and pollution of *Pfull*, then the optimum is pollution reduction of R^* and pollution of P^*.

Multiple Plants with Different Reduction Costs

Now, let's add two wrinkles to the basic model. First, suppose there are now two steel plants doing the dumping, with each plant dumping 200 units of sludge into the river each day. The marginal damage done by each unit of sludge is $100, as before. Second, suppose that technology is now available to reduce sludge associated with production, but this technology has different costs at the two different plants. For plant *A* reducing sludge is cheaper at any level of reduction, since it has a newer production process. For the second plant, *B*, reducing sludge is much more expensive for any level of reduction.

Figure 8 summarizes the market for pollution reduction in this case. In this figure, there are separate marginal cost curves for plant *A* (*MCA*) and for plant *B* (*MCB*). At every level of reduction, the marginal cost to plant *A* is lower than the marginal cost to plant *B*, since plant *A* has a newer and more efficient production process available. The total marginal cost of reduction in the market, the horizontal sum of these two curves, is *MCT*: for any total reduction in pollution, this curve indicates the cost of that reduction if it is distributed most efficiently across the two plants. For example, the total marginal cost of a reduction of 50 units is $0, since plant *A* can reduce 50 units for free; so the efficient combination is to have plant *A* do all the reducing. The socially efficient level of pollution reduction (and of pollution) is the intersection of this *MCT* curve with the marginal damage curve, *MD*, at point *Z*, indicating a reduction of 200 units (and pollution of 200 units)

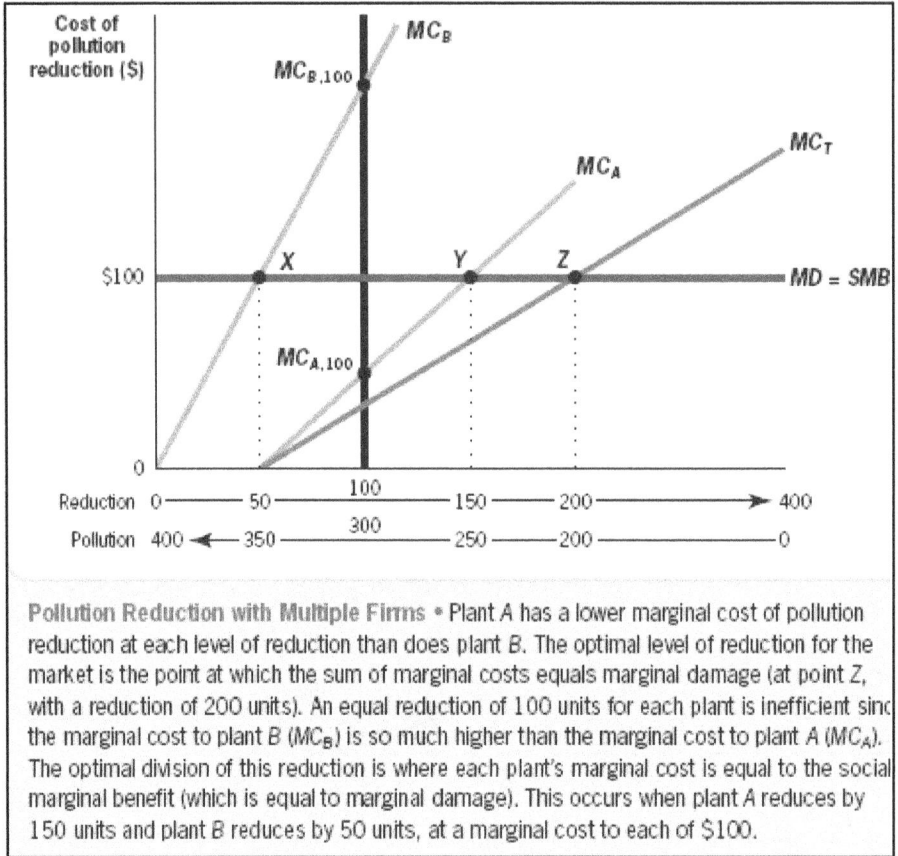

Pollution Reduction with Multiple Firms • Plant A has a lower marginal cost of pollution reduction at each level of reduction than does plant B. The optimal level of reduction for the market is the point at which the sum of marginal costs equals marginal damage (at point Z, with a reduction of 200 units). An equal reduction of 100 units for each plant is inefficient since the marginal cost to plant B (MC_B) is so much higher than the marginal cost to plant A (MC_A). The optimal division of this reduction is where each plant's marginal cost is equal to the social marginal benefit (which is equal to marginal damage). This occurs when plant A reduces by 150 units and plant B reduces by 50 units, at a marginal cost to each of $100.

Figure 8

Policy Option 1: Quantity Regulation Let's now examine the government's policy options within the context of this example. The first option is regulation: the government can demand a total reduction of 200 units of sludge from the market. The question then becomes: How does the government decide how much reduction to demand from each plant? The typical regulatory solution to this problem in the past was to ask the plants to split the burden: each plant reduces pollution by 100 units to get to the desired total reduction of 200 units.

This is not an efficient solution, however, because it ignores the fact that the plants have different marginal costs of pollution reduction. At an equal level of pollution reduction (and pollution), each unit of reduction costs less for plant A (MCA) than for plant B (MCB). If, instead,we got more reduction from plant A than from plant B,we could lower the total social costs of pollution reduction by taking advantage of reduction at the low-cost option (plant A). So society as a whole is worse off if plant A and plant B have to make equal reduction than if they share the reduction burden more efficiently.

This point is illustrated in Figure 8. The efficient solution is one where, for each plant, the marginal cost of reducing pollution is set equal to the social marginal benefit of that reduction; that is, where each plant's marginal cost curve intersects with the marginal benefit curve. This occurs at a reduction of 50 units for plant B (point X), and 150 units for plant A (point Y). Thus, mandating a reduction of 100 units from each plant is inefficient; total costs of achieving a reduction of 200 units will be lower if plant A reduces by a larger amount.

Policy Option 2: Price Regulation Through a Corrective Tax: The second approach is to use a Pigouvian corrective tax, set equal to the marginal damage, so each plant would face a tax of $100 on each unit of sludge dumped. Faced with this tax, what will each plant do? For plant A, any unit of sludge reduction up to 150 units costs less than $100, so plant A will reduce its pollution by 150 units. For plant B, any unit of sludge reduction up to 50 units costs less than $100, so it will reduce pollution by 50 units. Note that these are exactly the efficient levels of reduction! Just as in our earlier analysis, Pigouvian taxes cause efficient production by raising the cost of the input by the size of its external damage, thereby raising private marginal costs to social marginal costs. Taxes are preferred to quantity regulation, with an equal distribution of reductions across the plants, because taxes give plants more flexibility in choosing their optimal amount of reduction, allowing them to choose the efficient level.

Policy Option 3: Quantity Regulation with Tradable Permits: Does this mean that taxes *always* dominate quantity regulation with multiple plants? Not necessarily. If the government had mandated the appropriate reduction from each plant (150 units from A and 50 units from B), then quantity regulation would have achieved the same outcome as the tax. Such a solution would, however, require much more information. Instead of just knowing the marginal damage and the total marginal cost, the government would also have to know the marginal cost curves of each individual plant. Such detailed information would be hard to obtain.

Quantity regulation can be rescued, however, by adding a key flexibility: issue permits that allow a certain amount of pollution and let the plants trade. Suppose the government announces the following system: it will issue 200 permits that entitle the bearer to produce one unit of pollution. It will initially provide 100 permits to each plant. Thus, in the absence of trading, each plant would be allowed to produce only 100 units of sludge, which would in turn require each plant to reduce its pollution by half (the inefficient solution previously described).

If the government allows the plants to trade these permits to each other, however, plant B would have an interest in buying permits from plant A. For plant B, reducing sludge by 100 units costs $MCB,100$, a marginal cost much greater than plant A's marginal cost of reducing pollution by 100 units, which is $MCA,100$. Thus, plants A and B can be made better off if plant B buys a permit from plant A for some amount between $MCA,100$ and $MCB,100$, so that plant B would pollute 101 units (reducing only 99 units) and plant A would pollute 99 units (reducing 101 units). This transaction is beneficial for plant B because as long as the cost of a permit is below $MCB,100$, plant B pays less than the amount it would cost plant B to reduce the pollution on its own. The trade is beneficial for plant A as long as it receives for a permit at least MCA,100, since it can reduce the sludge for a cost of only MCA,100, and make money on the difference.

By the same logic, a trade would be beneficial for a second permit, so that plant B could reduce sludge by only 98, and plant A would reduce by 102. In fact, any trade will be beneficial until plant B is reducing by 50 units and plant A is reducing by 150 units. At that point, the marginal costs of reduction across the two producers are equal (to $100), so that there are no more gains from trading permits.

What is going on here? We have simply returned to the intuition of the Coasian solution: we have *internalized the externality by providing property rights to pollution.* So, like Pigouvian taxes, trading allows the market to incorporate differences in the cost of pollution reduction across firms.

Uncertainty About Costs of Reduction

Differences in reduction costs across firms are not the only reason that taxes or regulation might be preferred. Another reason is that the costs or benefits of regulation could be uncertain. Consider two extreme examples of externalities: global warming and nuclear leakage. Figure 9 extends the pollution reduction framework from Figure 7 to the situation in which the marginal damage (which is equal to the marginal social benefit of pollution reduction) is now no longer constant, but falling. That is, the benefit of the first unit of pollution reduction is quite high, but once the production process is relatively pollution-free, additional reductions are less important (that is, there are diminishing marginal returns to reduction).

Panel (a) of Figure 9 considers the case of global warming. In this case, the exact amount of pollution reduction is not so critical for the environment.

Since what determines the extent of global warming is the total accumulated stock of carbon dioxide in the air, which accumulates over many years from sources all over the world, even fairly large shifts in carbon dioxide pollution in one country today will have little impact on global warming. In that case, we say that the social marginal benefit curve (which is equal to the marginal dam- age from global warming) is *very flat:* that is, there is little benefit to society from modest additional reductions in carbon dioxide emissions.

Panel (b) of Figure 9 considers the case of radiation leakage from a nuclear power plant. In this case, a very small difference in the amount of nuclear leakage can make a huge difference in terms of lives saved. Indeed, it is possible that the marginal damage curve (which is once again equal to the marginal social benefits of pollution reduction) for nuclear leakage is almost vertical, with each reduction in leakage being equally important in terms of saving lives. Thus, the social marginal benefit curve in this case is *very steep.*

Now, in both cases, imagine that we don't know the true costs of pollution reduction on the part of firms or individuals. The government's best guess is that the true marginal cost of pollution reduction is represented by curve MC_1 in both panels. There is a chance, however, that the marginal cost of pollution reduction could be much higher, as represented by the curve MC_2. This uncertainty could arise because

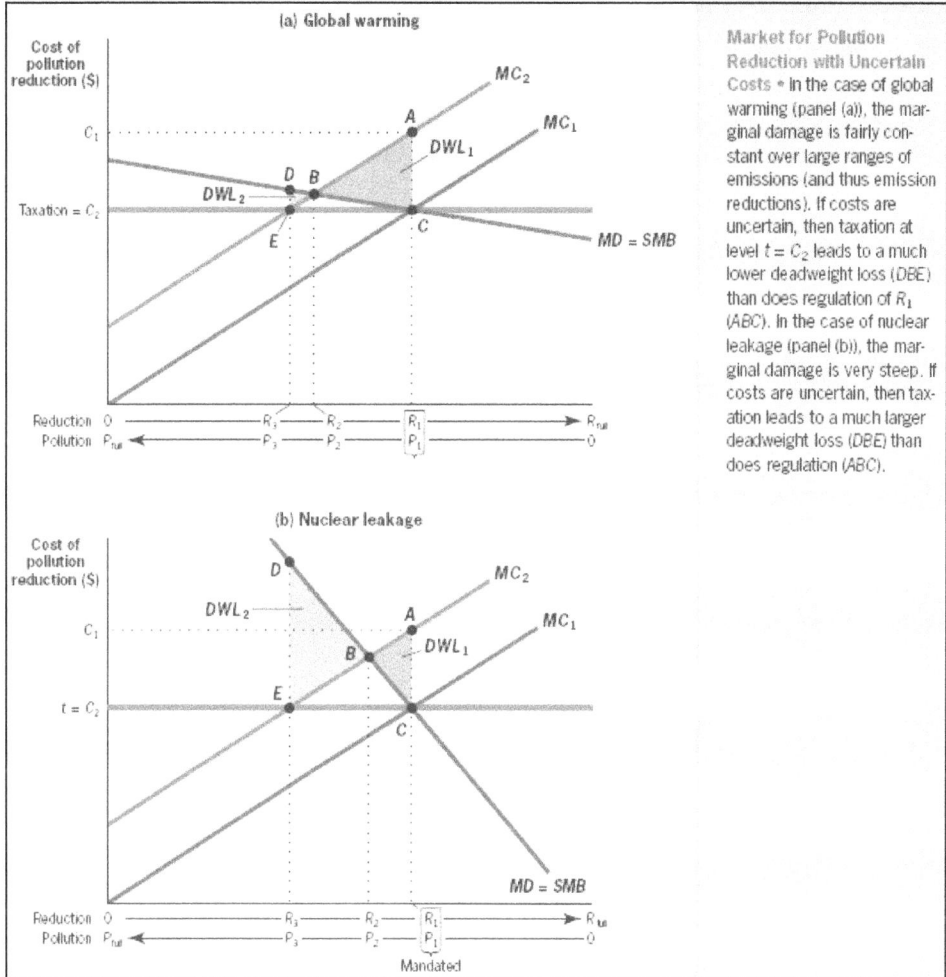

Figure 9

the government has an imperfect understanding of the costs of pollution reduction to the firm, or it could arise because both the government and the firms are uncertain about the ultimate costs of pollution reduction.

Implications for Effect of Price and Quantity Interventions

This uncertainty over costs has important implications for the type of intervention that reduces pollution most efficiently in each of these cases. Consider regulation first. Suppose that the government mandates a reduction, R_1, which is the optimum if costs turn out to be given by MC_1: this is where social marginal benefits equal social marginal costs of reduction if marginal cost equals MC_1. Suppose now that the marginal costs actually turn out to be MC_2, so that the optimal reduction should instead be R_2, where $SMB = MC_2$. That is, regulation is mandating a reduction in pollution that is too large, with the marginal benefits of the reduction being below the marginal costs. What are the efficiency implications of this mistake?

In the case of global warming (panel (a)), these efficiency costs are quite high. With a mandated reduction of R_1, firms will face a cost of reduction of C_1, the cost of reducing by amount R_1 if marginal costs are described by MC_2. The social marginal benefit of reduction of R_1 is equal to C_2, the point where R_1 intersects the SMB curve. Since the cost to firms (C_1) is so much higher than the benefit of reduction (C_2), there is a large deadweight loss (DWL_1) of area ABC (the triangle that incorporates all units where cost of reduction exceeds benefits of reduction).

In the case of nuclear leakage (panel (b)), the costs of regulation are very low. Once again, with a mandated reduction of R_1, firms will face a cost of reduction of C_1, the cost of reducing by amount R_1 if marginal costs are described by MC_2. The social marginal benefit of reduction at R_1 is once again equal to C_2. In this case, however, the associated deadweight loss triangle ABC (DWL_1) is much smaller than in panel (a), so the inefficiency from regulation is much lower.

Now, contrast the use of corrective taxation in these two markets. Suppose that the government levies a tax designed to achieve the optimal level of reduction if marginal costs are described in both cases by MC_1, which is R_1. As discussed earlier, the way to do this is to choose a tax level, t, such that the firm chooses a reduction of R_1. In both panels, the tax level that will cause firms to choose reduction R_1 is a tax equal to C_2, where MC_1 intersects MD. A tax of this amount would cause firms to do exactly R_1 worth of reduction, if marginal costs are truly determined by MC_1.

If the true marginal cost ends up being MC_2, however, the tax causes firms to choose a reduction of R_3, where their true marginal cost is equal to the tax (where $t = MC_2$ at point E), so that there is *too little* reduction. In the case of global warming in panel (a), the deadweight loss (DWL_2) from reducing by R_3 instead of R_2 is only the small area DBE, representing the units where social marginal benefits exceed social marginal costs. In the case of nuclear leakage in panel (b), however, the deadweight loss (DWL_2) from reducing by R_3 instead of R_2 is a much larger area, DBE, once again representing the units where social marginal benefits exceed social marginal costs.

Chapter 4

Market Failure Analysis

Market failure analysis is a type of economic analysis. It is important because it can indicate whether there is any prospect of a net benefit emerging form regulatory intervention. In particular, if regulatory intervention takes place when there is no market or regulatory failure, then that intervention will always impose net economic costs, no matter how carefully designed.

For this reason regulators are beginning to espouse market and regulatory failure analysis as a tool early in the policy formulation process to help them to determine whether there is a realistic prospect of a net economic benefit from regulatory action. We welcome these high-level commitments as marking clear progress in the broader cause of 'better regulation'. However, much remains to be done to build clarity on what is meant by 'market failure' within the context of a proper burden of proper burden of proof for intervention and the purpose of this note is to set out preliminary views on this question.

What is Meant by 'Market Failure'

In considering market failure form a regulatory perspective, it is important to draw a distinction between the theoretical notion of market failure, which is helpful, as an analytical tool, and a persistent 'welfare loss' arising form significant market failures that market forces are unable to resolve.

At the extreme, market failures arise when there are departures from economists' notion of a perfectly efficient market. In an efficient market firms produce goods and services at the lowest possible cost in terms of resources used and consumers buy the goods and services they want at the minimum possible price for a given quality. Moreover, at this price, supply and demand are in balance. To the extent that market failures arise, there is a waste of resources known as a 'welfare loss'.

In reality, all markets at any point in time exhibit some departures form the perfect market. Indeed, these departures are an essential feature of all dynamic and growing markets—the welfare loss creates profit making opportunities for market participants and in the vast majority of cases market participants can be expected to respond accordingly. It is only when this doesn't happen and the welfare loss is persistent and unresolved that one has a potential 'market failure case' for regulatory intervention.

In other words, the fact that a perfect market is unlikely to exist should not be a cause for regulatory concern: short-term welfare loss is not in itself a candidate for regulatory intervention. It only becomes a candidate if this welfare loss persists-market solutions are absent or partially ineffective.

Even then, regulation is unjustified if it does not improve the market outcome. As Callum McCarthey, Chairman of the Financial Services Authority has said.

'........regulatory action should only be taken when there is a market failure. New this is in fact a weak definition of the circumstances of when regulatory action is justified, since all realistic markets – that is all markets which exist in practice – have some elements of market failure. It is an argument too often deployed by those who fovour intervention that any market failure justifies intervention. The strong – and to me correct-test goes beyond that: there be both market failure and the prospect that intervention will provide a net benefit. This involves recognizing that regulatory intervention has a cost and. a probability of failure. Identification of a market failure should not lead to the assumption that regulatory failure is less likely, or less costly. It is an open and empirical question, which needs analysis on a case by case basis.'

Types of Market Failure

Central to an analytical understanding of market failure is the point that when markets are working well the price of any good or service will equal both the marginal cost and the marginal benefit of that product.

For analytical purposes there are three types of market failure that matter most in the context of financial markets. All of which involve sustained departures from the price equals marginal cost equals marginal benefit paradigm:

- ☆ buyers (or sellers) are not able to from reasonably accurate estimates of actual marginal benefit (marginal cost) leading to 'wrong' consumption (or production) decisions – imperfect or asymmetric information;
- ☆ 'social' marginal benefit or cost differs from the 'private' marginal benefit or cost leading to under or over provision of a good or service-externalities;
- ☆ market power on the part of seller(s) (or buyer(s)) means that price exceeds marginal benefit (marginal cost), leading to 'excess profits' ('excess consumer surplus') and under-provision of a good or service-monopolies/ oligopolies (monopsonies).

Imperfect and Asymmetric Information

Individual decisions are affected by imperfect information about quality, price and the future. Information is a source of market failure if it is understandable by, useful to but not available to the buyer or seller, despite willingness on the part of the buyer or seller to pay for the costs of producing that information. Importantly, in this context 'useful' means information that could change selling or buying behaviour as distinct from information that is a nice 'optional extra'.

Externalities

A good or service generates externalities if its production or consumption affects the welfare of people or firms other than its original producers or consumers without prices reflecting such effects. Externalities may be negative or positive. They are 'negative' for those on whom they impose costs and 'positive' for those who gain from them. Negative externalities occur when production decisions do not take account of all the costs flowing form that decision because these are costs not borne by the firm. The classic non-financial sector example is pollution from factories. The classic example of a negative externality in financial services is systemic risk, where the failure of one bank may lead to runs on other banks and hence to problems for those other banks and their customers.

Positive externalities arise when the decision to consume does not take account of all of the costs of production. If this market failure is sustained, production is too low. The classic non-financial example is the production of television signals which, in theory, can be enjoyed freely by any consumer with a television set. Financial sector examples can in practice be those induced by regulation (leading to regulatory failure-more on this below). For example, in the bond market transparency debate a wrong diagnosis and/or mistreatment of an information asymmetry could in fact create externalities that lead to market shrinkage.

'If information gathering is costly, or valuable knowledge arises as a consequence of undertaking a costly activity (*e.g.*, taking an own account position) then differences in trading information may efficiently continue to exist if the person who might benefit form that information does not perceive the benefit as sufficient to compensate the person generating that underlies the potential for a negative relationship between mandated transparency and loss of liquidity relates to this point. In particular, mandated transparency in the above situation, which in this case might be the incentive to maintain the provision of quotes in particular markets.

Market Power

Market power on the sell-side refers to the situation where revenues above the marginal cost of all production input, including the cost of capital, ('excess profits'), can persist rather than be eroded by competitive pressures. 'Excess profits' are also an expected outcome of the market failure of information asymmetry, in that an information asymmetry reduces the buy-side ability to put effective competitive pressure on sell-side.

A good example of how this analysis can aid 'market failure' decisions is found in the EU bond market transparency debate. An important observation in the analysis of the relevant European markets is the tightness of the bid-ask spreads (both quotes and actual trades). This finding suggests that the relevant markets are working well-neither the market failure of 'market power'/ excess profit' nor the market failure of persistent information asymmetries (warranting transparency regulation) would appear to be consistent with this finding.

The causes of sell-side market power can be either natural or induced. A natural monopoly / oligopoly will arise where the magnitude of the sunk costs acts as a barrier to entry, so reducing the degree of competition. Classic examples are the provision of clean water and railways. Artificial monopolies can result from public sector intervention, either in the form of raising entry hurdles for particular competitors competitions (protectionism) or all potential competitors equally (*e.g.*, 'fit and proper' conduct of business requirements).

Regulatory Failure

'Regulatory failure' itself is used as an economic justification for intervention. For example, the economic justification for MiFID was the prevalence of public policy barriers preventing the emergence of a single market for financial services.

More generally, market failure and regulatory failure are closely related considerations. Regulatory interventions generally do increase the cost of producing financial services and often, through conduct of business and prudential requirements, create barriers to entry which reduce competition, increase costs to consumers and lead to under production. Consequently, any regulatory intervention which does not respond to a market failure or does not respond proportionately to a market failure, or 'crowds out' a market forces response to a market failure will impose net economic costs.

In order to avoid an intervention whose economic costs are higher than expected, or whose benefits are lower than expected, it is important to ensure that the market failure analysis includes the following two stages:

☆ assessment of the nature and magnitude of the market failure, using data wherever possible; and

☆ analysis of whether or not market failures are likely to be corrected by market forces.

Evidence of significant market failures might include:

☆ a wide dispersion of market prices for essentially the same product (including risk profile);

☆ persistent 'excess profits'; and

☆ persistent mismatch between risk and return.

Even when there is a significant market failure, factors weighing in favour of no intervention include:

☆ evidence that the market failure has arisen only recently;

☆ the recent introduction of other regulation likely to mitigate the market failure;

☆ evidence that the market is able to discern and respond to the market failure – for example new entrants, and technology.

Market power causes the price to move above the social optimum price and quantity to fall below the social optimal quantity. As a result both producers and consumers experience welfare loss. The deadweight welfare loss deadweight welfare loss can be measured in terms of the loss in consumer and producer surplus due to the lower quantity and higher prices resulting from market power.

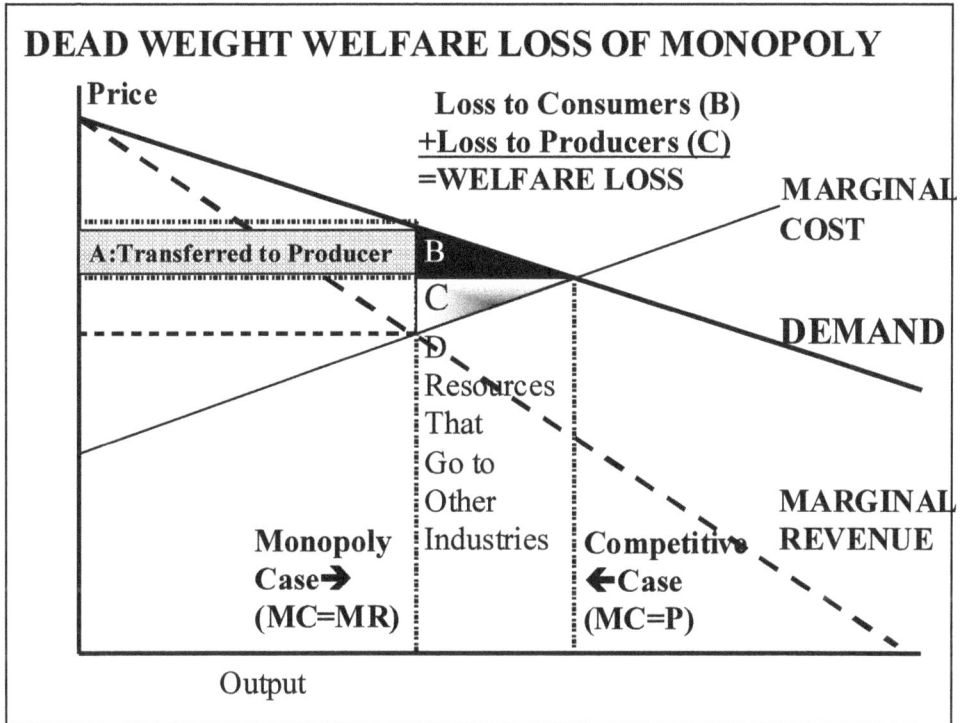

DEAD WEIGHT WELFARE LOSS OF MONOPOLY

Figure 1

NOTE: With monopoly the consumer loses the consumer surplus (areas, A and B). However, the producer gains only part of this consumer surplus (area A); area B is lost. In addition the producer loses some producer surplus, area C. The shaded region (area B and C) is the deadweight loss.

By intervening with antitrust policy, the government hopes to lower prices and increase quantity. However, antitrust policies may not effectively correct structural problems leading to market power.

Figure 1 which shows the hypothetical demand curve facing a monopolist. We can identify the output at which the monopolist maximizes profit (marginal revenue=marginal cost) and the competitive equilibrium (price=marginal cost). By raising its price the monopolist can grab some consumer surplus (area "A") which offsets its loss in producer surplus (area "C"). However, higher prices reduce output which reduces benefits to both consumers and producers. Much of this loss is compensated by sending the resources to offsetting uses elsewhere in the economy

(Area D in Figure 1). However, there is still a little triangle (Shaded areas B and C) which is lost both to consumers and producers from the cutback in production. No one receives the advantage of this reduction and it is therefore referred to as the deadweight welfare loss of monopoly pricing.

Externalities

A social optimum social optimum should occur where net social benefits are maximized. The net social benefit The net social benefit is the difference between social benefits and social costs. The concept of social benefit in the public sector is parallel to the concept of total revenue in the private sector. Similarly the concept of social cost is parallel to the concept of total cost in the private sector. However, benefits or costs accruing to anyone and everyone, not just the stockholders, must be included in the measurement of net social benefits. As we will see in section 3, such a broad accounting requires more creativity and more measurement problems than a firm would face.

Table 1: Social Benefits and Costs of Education

No. of Students (millions per year) (1)	Total Benefit ($ millions per year) (2)	Total Cost ($ millions per year) (3)	Marginal Social Benefit ($ per student) (4)	Marginal Social Cost ($ per student) (5)
0	0	0	–	–
1	20,000	1000	20000	1000
2	35,000	6000	15000	6000
3	45,000	16000	10000	10000
4	50,000	36000	5000	20000

Note: The marginal benefit is the difference of the total benefit from column 2 divided by the increment in the number of people who receive the benefit from column 1. Marginal costs are computed similarly.

The problem of finding maximum net social benefit is parallel to the problem of maximizing profit. In fact net social benefit is often referred to as social profit. Marginal social benefit (MSB) **Marginal social benefit** (MSB) is the amount of social benefit from the next extra unit of a good or service. Marginal social cost (MSC) **Marginal social cost** (MSC) is the amount of social cost from producing the next extra unit. The first order condition for a maximum requires that the two be equal:

$$MSB = MSC$$

This first order condition is quite similar to the condition that marginal revenue equal marginal cost that applied to the private sector.

Suppose Table 1 and Figure 2 depict the social benefits and costs of education in the U.S. The social benefit curve suggests diminishing returns to greater education; each successive group of students returns a little less in social benefits to society (column 4 of Table 1 is declining). The social cost curve suggests increasing costs to

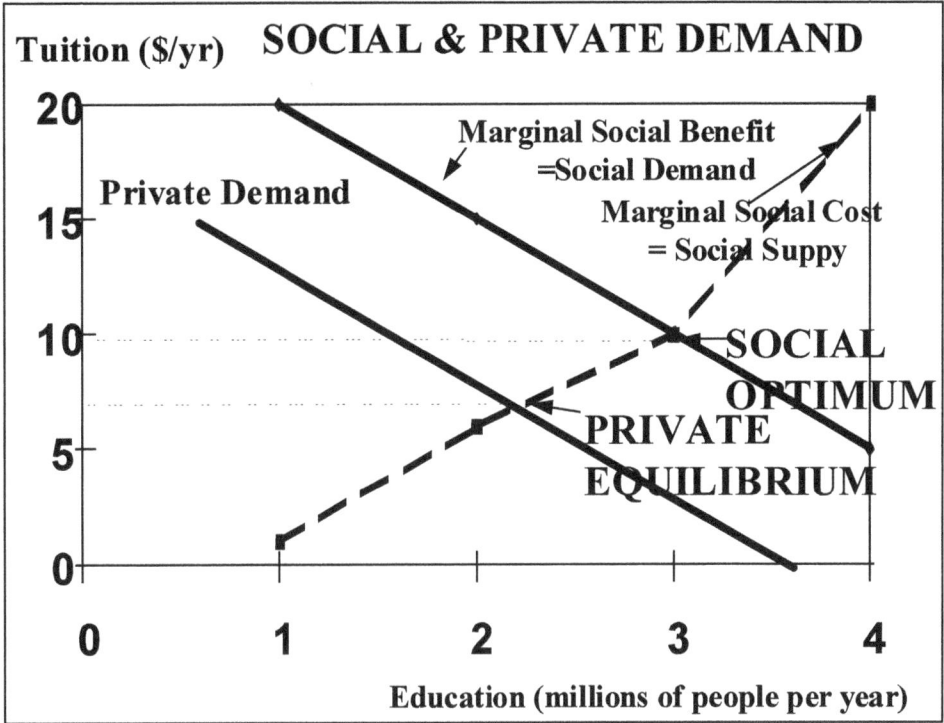

Figure 2

NOTE: When there are positive consumption externalities the social demand curve will be above the private market demand curve.

greater education; as more people are drawn from other jobs into the education system, society experiences greater and greater opportunity costs (column 5 is rising).

The marginal social benefit curve reflects what society is willing and able to buy at different prices and it therefore is effectively the same thing as the social demand curvesocial demand curve. When the market demand curve also corresponds to the social demand curve, it is easy to measure who receives the benefits. Each point of the demand curve represents the benefit, referred to as the marginal valuemarginal value, received by the next individual member of society. The area under the demand curve corresponds to the total benefit that consumers receive from a given amount of output. The difference between the total benefit and the price consumers pay for the output is called the consumer's surplusconsumer's surplus. Similarly the area between the supply curve and the equilibrium price goes to producers and is often referred to as the producer's surplusproducer's surplus. The producer's surplus corresponds to the amount of revenue received by the producers minus the cost of providing the output.

Society may value consumption differently than an individual consumer. Generally, if there are parties other than the consumers of a product who are affected by the consumption of the product, then the private and social demand curves diverge. The difference between the social private and social demand curves are a measure of the consumption externalitiesconsumption externalities from the consumption of the good. If the externalities benefit third parties – besides the producer or consumer– then the social demand curve will be above the private market demand curve. However, if the consumption externalities hurt third parties, then the social demand curve will be below. Each point on the social demand curve represents the marginal social valuemarginal social value of consumption.

Similarly society may value production differently than an individual producer. Generally, if there are parties other than the producers who are affected by the supply of a good or service, then the private and social cost curves diverge. The difference between the social private and social marginal cost (or social supply) social marginal cost (or social supply) curves are a measure of the production externalities production externalities from the supply of the good. If the externalities benefit third parties then the social marginal cost curve will be below the private marginal cost curve. However, if the production externalities hurt third parties, then the social marginal cost curve will be above.

The social optimum social optimum occurs where social demand and supply intersect. When market and social demand (or marginal cost) curves diverge, the social optimum is not the same as the private market equilibrium and the market will fail to achieve the social optimum.

There may be many reasons for society valuing an individual's education more than the individual alone does. Education may contribute to greater stability, growth potential, and social acceptability. while an individual may think getting educated just means getting a better paying job. Suppose society recognizes $5000 more of benefits than each person who receives the education. In Figure 2, the social demand curve would therefore be $5000 above the private market demand curve. Each point on the social demand curve would represent the marginal social value marginal social value of education. The difference between the private and social demand curves is the positive consumption externality consumption externality from education, measured as $5000 per person.

In Figure 2, the private market demand curve is below the social demand curve. As a result the private market equilibrium indicates less education at a lower price than would be desirable at the social optimum. The market fails to provide the socially desirable level of education. The government may have to intervene in the education market.

Externalities prevent the private market equilibrium from correctly delivering the social optimum. To correct such a market failure, the government can intervene in a number of ways:

1. *Prohibitions on production or consumption*: Prohibitions are suited to goods which produce clear public harm without compensatory public benefit. In the United States this kind of intervention is used for many types of drugs, other harmful products, prostitution, violence, and a wide number of antisocial behaviours.

2. *Nationalization or Public Enterprise*: When a good is highly dangerous but has a clear, offsetting, and overriding benefit, the government may nationalize nationalize a private firm or go into business itself as a public enterprisepublic enterprise. The government sets up public enterprises like NASA to run the space program, the U.S. Post Office, the public education system, and many other services.

3. *Regulation*: The government can directly limit prices, output, profit, conduct, and performance of a firm. The government closely regulates utilities, government contractors, the banking system, and firms engaged in national security.

4. *Taxes and Subsidies*: Taxes and subsidies can be tailored in such a way that they force a firm to pay the full cost (or receive the full benefit) of an externality. For example, government imposes specific taxes on cigarettes, alcohol, gambling and other slightly deleterious luxuries. On the other hand, it has subsidized semiconductors, synthetic fuels, American shipping, and other industries contributing to national security.

5. *Creating new markets or adjusting existing ones*: Often a market failure can be corrected by changing market incentives. For example, states like California have created pollution rights to streams or environments. If environmentalists wish to buy the pollution rights, they can and pollution will be prevented. However, if firms receive the highest bid, they can pollute according to the right that they purchase.

Generally the least intrusive or restrictive form of government intervention should be used to correct distortions due to externalities.

Divisibility and Excludability

A pure private good or service pure private good or service can be divided into infinitely small units for consumption and can be allocated strictly to the people who buy the good. A purely private good does not have impacts on parties other than the seller and buyer which means there are no externalities. By contrast a pure public good a pure public good cannot be divided into discrete quantities that each customer can buy, and it actually costs something to exclude some people from enjoying the good.

A good which does not have complete divisibility or excludability may be difficult for a private market to deliver at the social optimum. Without excludability excludability people have an incentive to let someone else pay and then become free riders free riders in the enjoyment of the good once it has been paid for. Frequently indivisible indivisible goods or services those which cannot be divided among consumers do not have complete excludability. We cannot subdivide a park, giving everyone in the city a square inch of it. Effectively there is a choice to have the park or not to have it; after deciding to have the park, everyone can use it and it would cost a great deal of money to try to exclude anyone.

The means of correcting the problems of indivisibility and non-excludability are similar to the means for correcting externalities that were examined above. One of the

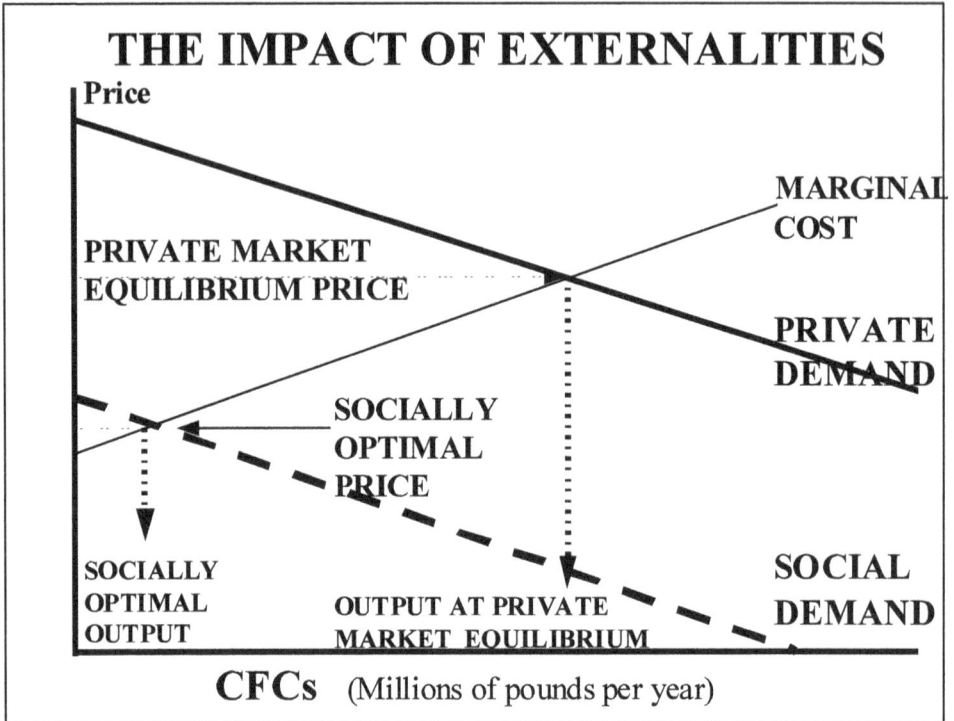

THE IMPACT OF EXTERNALITIES

Price

MARGINAL COST

PRIVATE MARKET EQUILIBRIUM PRICE

PRIVATE DEMAND

SOCIALLY OPTIMAL PRICE

SOCIALLY OPTIMAL OUTPUT

OUTPUT AT PRIVATE MARKET EQUILIBRIUM

SOCIAL DEMAND

CFCs (Millions of pounds per year)

Figure 3

NOTE: A negative consumption externality means that social demand is below the private market demand curve. Private market equilibrium is at a higher price and quantity than the social market optimum.

most potent weapons of our legal system is the legal procedure for ascribing blame and assessing damages from those responsible. For example, the problem of excludability becomes a problem with a public bad public bad like depletion of the ozone layer. There is no way to insulate anyone from the effects and it is costly and difficult to allocate the effects to the people who cause the problem.

Information

For firms to contest in a market, there must be information not only on the profitability of the market but upon the technology necessary to enter the market. A firm must be able to find knowledgeable personnel, know what resources to use, know how to use them, and know demand conditions in order to enter a market successfully. Information provides customers with the ability to bargain prices downward to the costs of providing a good. Information also allows sellers to find the best markets for their goods.

With imperfect access to information by either customers or buyers a market may not reach a social optimum. To assure that markets work efficiently the government may intervene to provide adequate information as it does in agriculture, utilities, the banking system, the business census, and labour.

Access to information does not mean that managers of a firm must know everything about a market, but only that they be able to find out what other managers in the market know or are able to know. Great opportunities to enter a market often occur precisely when there is little or no information about the market.

Dynamic Market Failure

Markets may be structured in such a way that they become dynamically unstable when left to themselves. Lags in obtaining information and responding effectively to it can lead to dynamic instability in a market. Such instability may cause a market to wander away from a social optimum. Following are three examples of dynamic market failures from such instability:

The Cobweb Model and Lagged Adjustment

When demand is relatively inelastic while supply is very elastic, lagged adjustment can move a market away from equilibrium as shown in the hypothetical chemical example pictured in Figure 4. A price above equilibrium one year causes chemical firms to supply a large quantity of goods the next year; in effect the firms are

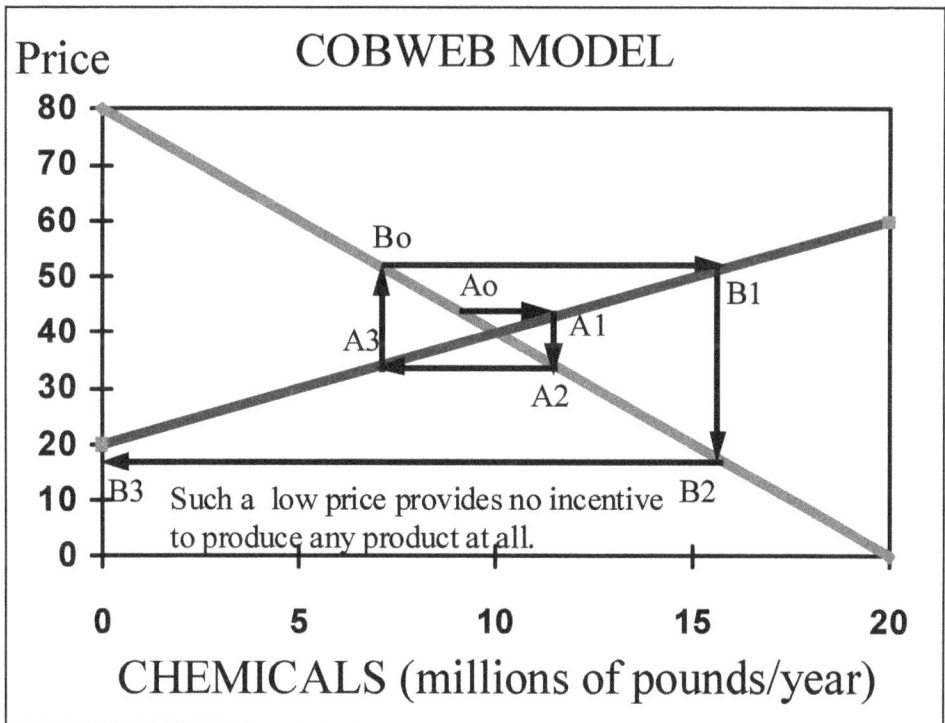

Figure 4

NOTE: In response to high prices at A_o Chemical companies would expand capacity to produce A_1. However, with excess capacity prices fall to A_2 and capacity exits to A_3. The shortage in capacity raises prices up to B_o. The cycle starts all over, but the swings are more dramatic leading to dynamic market failure.

responding with a one year lag to the price signals of the previous year. However, the larger quantity supplied results in a precipitous price reduction. Again with a one year lag firms respond by cutting back the quantity supplied. However, the lower quantity supplied results in skyrocketing prices. The process continues and progressively moves away from the equilibrium quantity and supply.

Overuse of Replenishable Resources

Many populations of plants and animals replenish themselves to produce a sustained yield for industry. However, if the rate of utilization reaches past a certain critical level, the density of the resource population may be reduced to a point where the population can no longer sustain itself. If firms do not know the critical level for a resource population or experience a lag in being able to measure the use of a resource, they may inadvertently reach past the critical point and send the resource into extinction. In this case lack of information or a lag in monitoring information is the source of the problem.

Even with knowledge about sustaining the yield of a population, market incentives may lead to the demise of the population. A prisoner's dilemma faces the firms using the population. If everyone cooperates, sustained yield can be achieved. If they don't cooperate the population will become extinct. However, if some individuals cooperate while others do not, the resource population will not only become extinct but the cooperating individuals will lose out to those who do not cooperate as the resource disappears.

Non-replenishable resourcesNon-replenishable resources. Even when resources are not replenishable there may be an optimum rate of exploitation based on the speed with which they can be replaced with substitutes. However, invention, innovation, and diffusion must work smoothly to produce the new substitutes. Otherwise society faces a crisis of market failure.

While market failure provides a basis for government intervention, such intervention has its own costs. The benefits of government intervention must be weighed against the costs of that intervention.

Government Failure

Implicitly government has been assumed so far to be capable of correcting market failure. However, government involvement may make a market failure worse rather than better. Furthermore, the government may set up rules which lead to market failure. Finally government itself requires resources which can place a heavy drain on an economy. In this section we'll examine these government failures from the point of view of a former CEO of the Dupont Corporation, named Irving Shapiro.

Externalities and Side Effects

Government policies can have their own externalities and side effects which may not always be desirable. When the Soviet Union fell apart in the early 1990s it became apparent that the Soviet government had permitted massive pollution and destruction of its territories. The American government itself has been a major polluter.

Oversight of the government by the government itself produces an obvious conflict of interest.

Economic Distortions

Government policy can also lead to inefficiencies. The government has been very slow to realize the sophisticated workings of markets. By imposing rules, a government agency can create surpluses or shortages. Either type of market disequilibrium results in waste.

Once again it is instructive to look at the example of the Soviet Union before it fell. As American director of the U.S.–U.S.S.R. Trade and Economic Council, Inc., Du Pont's Shapiro had become acquainted with a classic example of the inefficiencies of government:

> It's much worse in Russia, because the only structure in their society is government. All the VIPs are government officials. The bureaucracy, the rules, the regulations, the paperwork- why, it's simply overwhelming. To paraphrase an old Churchill line about democracy, "It's got a lot of defects; it just happens to be better than anything else we can come up with.'

By attempting such complete control, the Russian government also took the complete blame for what occurs to the economy. It is much easier to allow the private sector to handle the intricacies of the market and for government to step in whenever there are abuses. Socialist and Communist governments throughout Europe and the Eastern block began a retreat toward reliance more on the private sector. Specifically, many government industries are being sold to the private sector in a move called privatization.

Information

The government should attempt to make more information available to allow markets to work competitively and efficiently. However, under government auspices firms can often cooperate and prevent outsiders from getting information about the cooperative effort. For example, when firms get together to lobby for a particular price control ceiling, or target price for protectionism, they may exclude other participants in the market who would be hurt by the policy. Under the guise of national security, the government may withhold information about projects that it undertakes with firms; such insulation from public scrutiny prevents the forces of competition from controlling prices and costs. Ultimately whenever the government allows one firm in a market to gain a relative advantage in the acquisition of information, it prevents the competitive market from working efficiently.

Equity

Government must often define as well as find policies to achieve equity among the members of society. Typically the government tries to use taxes and subsidies as a means for correcting inequities. However, such policies may result in their own inequities.

When the government attempts to make transfers itself by taxing on the one hand and subsidizing on the other it faces substantial incentive problems. First of all, the bureaucracy needed to administer such transfers is expensive and adds to the government deficit. Secondly, the government finds it easier to disperse money than to collect it which means there tends to be a bias for the government to run a deficit. Thirdly, the government's policies are likely to decrease private market incentives. When the government taxes it inevitably penalizes earners. Implicit in any tax policy is a marginal rate of taxation marginal rate of taxation which indicates how much of every extra dollar earned is actually received as after tax income. Instead of receiving a dollar for every dollar earned, individuals may receive only eighty cents which means there is a marginal tax rate of 20 per cent. High marginal taxation lessens the incentive for people to earn money. Governments therefore face a trade off between equitable transfers and efficient incentives.

Of course, the government can avoid taxation by deficit financingdeficit financing. But deficits cause the U.S. debt to stack up which scares business. CEO Shapiro, whom we have already quoted above, stated:

> ...... The heart of the problem is a political system in which the people in public office want to promise every constituent group whatever they want and our society can't produce enough wealth to do everything. But this can be overcome by good management. It's no different than running Du Pont. I can't invest more capital than I have, and I know that. We have to tailor our investment programme to our capital and our borrowing capacity. And in government, you can't give away more than you've got without creating inflation and other serious consequences. As soon as the American people accept that premise and start putting heat on the politicians as, for instance, the Germans have done, then we can lick inflation

The analogy between the government and firms fails to account for the government's ability to tax. Nevertheless, the fact that so many experienced business leaders believe in this analogy whether right or wrong means that larger government deficits trigger fears of inflation, higher interest rates, and heavy debt burdens for future generations. These fears surface in market reactions to government financial data and attempts by the business community to reform government.

Lags and Dynamic Government Failure

As in the private market, lags can play a destabilizing role in the government's interaction with markets. There are four basic lags between the time that a problem arises and the time that the government does something about it:

1. *Recognition lag*: Between the time a problem occurs and the time it is recognized as a problem by a government there is a recognition lag recognition lag.

2. *Response lag*: Between the time a problem is recognized and the time that a decision is made about what to do there is a response lagresponse lag.

3. *Implementation lag*: Between the time a problem is responded to and the time that action is taken, there is an implementation lag implementation lag.

4. *Impact lag*: Between the time of implementation and the time when the final impact is felt, there is an impact lag impact lag.

If the pollution problem of ozone were to progressively deteriorate the above schedule could be far too slow to take care of the problem. The government would always be responding with too little action, too late and there would be clear government failure.

Chapter 5

Environment Kuznets Curve

The EKC says that the pollution will first increase with the level of GDP per capita, reach maximum at around $8,000 and then decrease at higher levels of income. The policy implications of this finding according to some are grow first and then clean up. Some have argued that economic growth is a panacea or "cure all" for environmental degradation, "in the end the best and probably the only-way to attain a decent level of environment quality." Another writer claims that existing environmental regulations by reducing growth may actually be reducing environmental quality.

Explanations for Environmental Kuznets Curve

1. A natural progression of economic development from clean agrarian economies to polluting industries to clean service economies.
2. Advanced economies exporting their pollution to less developed countries.
3. The internalization of externalities requires relatively advanced institutions for collective decision-making.
4. Another model is that below a threshold level of pollution only the dirtiest technology will be used.
5. Environmental quality is a stock resource that degrades over time.
6. Demand for environmental quality overtakes supply ultimately.
7. Decreasing costs in pollution abatement.

One of the important implications of an environmental Kuznets curve (EKC) is that growth and development in a country need not lead to environmental degradation.

One explanation for the environmental Kuznets curve is that the income elasticity of marginal damage is increasing in income. So, at low levels of income, pollution will rise with neutral growth because the policy response is weak. As income rises, the policy response becomes stronger, and if at some point the income elasticity of marginal demand is sufficiently high, pollution will start to fall as income increases. Any theory of the EKC requires some force to eventually more than fully offset the scale effect of growth. In the income-effect explanation it is primarily a technique effect that does this. At low incomes, pollution initially rises with growth because increased consumption is valued highly relative to environmental quality. As income rises, the willingness to pay for environmental quality rises, and increasingly large sacrifices in consumption are made to provide great environmental benefits.

Pessimists have argued the cross-section evidence is nothing more than a snapshot of a dynamic process. Also, that globalization promotes a race to the bottom. Furthermore, while certain pollutants decrease when income increase industrial society continuously creates new, unregulated, potentially toxic pollutants. The stakes are high in this debate as those who believe in race to the bottom argue in favour of trade and investment restrictions to eliminate the cost of pollution "havens."

Dasgupta and his co-authors in the article that you have given an optimistic version of the EKC. They believe that the EKC can be lower and flatter. The reasons they give include

1. Environmental regulation.
2. Economic liberalization (compositional effects of trade and scale economies)
3. Informal regulation.
4. Pressure from market agents.
5. Better methods of environmental regulation.
6. Better information.

You should be able to provide details on these.

There is empirical evidence that the amount of environmental regulation increases with the level of income. The reasons given are the standard ones:

1. Pollution damages gets higher priority after society has competed investments in health and education.
2. High-income societies have more plentiful personal and budgets for monitoring enforcements.
3. Higher income and education empower local communities to enforce higher environmental standards.

Theoretical Background

The EKC concept emerged in the early 1990s with Grossman and Krueger's (1991) pathbreaking study of the potential impacts of NAFTA and Shafik and Bandyopadhyay's (1992) background study for the 1992 World Development Report. However, the idea that economic growth is necessary in order for environmental quality to be maintained or improved is an essential part of the sustainable development argument promulgated by the World Commission on Environment and

Development (1987) in *Our Common Future*. The EKC theme was popularized by the World Bank's *World Development Report 1992* (IBRD, 1992), which argued that: "The view that greater economic activity inevitably hurts the environment is based on static assumptions about technology, tastes and environmental investments" (p. 38) and that "As incomes rise, the demand for improvements in environmental quality will increase, as will the resources available for investment" (p. 39). Others have expounded this position even more forcefully with Beckerman (1992, p. 482) claiming that "there is clear evidence that, although economic growth usually leads to environmental degradation in the early stages of the process, in the end the best – and probably the only – way to attain a decent environment in most countries is to become rich." However, the EKC has never been shown to apply to all pollutants or environmental impacts and recent evidence (Dasgupta *et al.*, 2002; Perman and Stern, 2003) challenges the notion of the EKC in general. The remainder of this section discusses the economic factors that drive changes in environmental impacts and may be responsible for rising or declining environmental degradation over the course of economic development.

If there were no change in the structure or technology of the economy, pure growth in the scale of the economy would result in a proportional growth in pollution and other environmental impacts. This is called the scale effect. The traditional view that economic development and environmental quality are conflicting goals reflects the scale effect alone. Proponents of the EKC hypothesis argue that "at higher levels of development, structural change towards information-intensive industries and services, coupled with increased environmental awareness, enforcement of environmental regulations, better technology and higher environmental expenditures, result in leveling off and gradual decline of environmental degradation." (Panayotou, 1993, p. 1).

Therefore, at one level the EKC is explained by the following 'proximate factors':

1. Scale of production implies expanding production at given factor-input ratios, output mix, and state of technology.

2. Different industries have different pollution intensities and typically, over the course of economic development the output mix changes.

3. Changes in input mix involve the substitution of less environmentally damaging inputs for more damaging inputs and vice versa.

4. Improvements in the state of technology involve changes in both:

   a) Production efficiency in terms of using less, *ceteris paribus*, of the polluting inputs per unit of output.

   b) Emissions specific changes in process result in less pollutant being emitted per unit of input.

These proximate variables may in turn be driven by changes in underlying variables such as environmental regulation, awareness, and education in the course of economic development. A number of papers have developed theoretical models about how preferences and technology might interact to result in different time paths of environmental quality. The different studies make different simplifying assumptions about the economy. Most of these studies can generate an inverted U shape curve of pollution intensity but there is no inevitability about this. The result depends on the assumptions made and the value of particular parameters. Lopez (1994) and Selden and Song (1995) assume infinitely lived agents, exogenous technological change and that pollution is generated by production and not by consumption. John and Pecchenino (1994), John *et al.* (1995), and McConnell (1997) develop models based on overlapping generations where pollution is generated by consumption rather than by production activities. Stokey (1998) allows endogenous technical change. It seems fairly easy to develop models that generate EKCs under appropriate assumptions but none of these theoretical models has been empirically tested. Furthermore, if in fact the EKC for emissions is monotonic as more recent evidence suggests, the ability of a model to produce an inverted Ushaped curve is not necessarily a desirable property.

Econometric Framework

The earliest EKCs were simple quadratic functions of the levels of income. However, economic activity inevitably implies the use of resources and by the laws of thermodynamics, use of resources inevitably implies the production of waste. Regressions that allow levels of indicators to become zero or negative are inappropriate except in the case of deforestation where afforestation can occur. This restriction can be applied by using a logarithmic dependent variable. The standard EKC regression model is:

$$\ln(E/P)_{it} = \alpha_i + \gamma_t + \beta_1 \ln(GDP/P)_{it} + \beta_2 (\ln(GDP/P))^2_{it} + \varepsilon_{it} \qquad (1)$$

where E is emissions, P is population, and ln indicates natural logarithms. The first two terms on the RHS are intercept parameters which vary across countries or regions i and years t. The assumption is that though the level of emissions per capita may differ over countries at any particular income level the income elasticity is the same in all countries at a given income level. The time specific intercepts are intended to account for time varying omitted variables and stochastic shocks that are common to all countries.

Usually the model is estimated with panel data. Most studies attempt to estimate both the fixed effects and random effects models. The fixed effects model treats the αi and γt as regression parameters while the random effects model treats them as components of the random disturbance. If the effects αi and γt and the explanatory variables are correlated, then the random effects model cannot be estimated consistently (Mundlak, 1978; Hsiao, 1986). Only the fixed effects model can be estimated consistently. A Hausman (1978) test can be used to test for inconsistency in the random effects estimate by comparing the fixed effects and random effects slope parameters. A significant difference indicates that the random effects model is estimated inconsistently, due to correlation between the explanatory variables and the error components. Assuming that there are no other statistical problems, the fixed effects model can be estimated consistently, but the estimated parameters are conditional on the country and time effects in the selected sample of data (Hsiao, 1986). Therefore, they cannot be used to extrapolate to other samples of data. This means that an EKC estimated with fixed effects using only developed country data might say little about the future behavior of developing countries. Many studies compute the Hausman statistic and finding that the random effects model cannot be consistently estimated estimate the fixed effects model. But few have pondered the deeper implications of the failure of this orthogonality test.

Perman and Stern (2003) employ some newly developed panel unit root and cointegration tests and find that sulfur emissions and GDP per capita may be integrated variables. Coondoo and Dinda (2002) yield similar results for carbon dioxide emissions. If EKC regressions do not cointegrate then the estimates will be spurious. Very few studies have reported any diagnostic statistics for integration of the variables or cointegration of the regressions and so it is unclear what we can infer from the majority of EKC studies.

Functional Forms for the Environmental Kuznets Curve

The Environmental Kuznets Curve hypothesis is easily formulated in an IPAT framework. In IPAT, an environmental impact I is expressed as a product of factors – population, P, "affluence" (GDP per capita), A and "technology," T:

$$I = P \times A \times T$$

In this equation, because P multiplied by A is just GDP, the technology factor T must be the ratio of the environmental impact to GDP. The Environmental Kuznets Curve hypothesis is a statement about T. It says that T is a function of A that increases at small values of A and declines at high values of A. Any functional form for an environmental Kuznets curve must reflect this.

Regression Equations

In empirical studies of the Environmental Kuznets Curve hypothesis, a particular environmental impact (or an impact indicator) is chosen. For example, as an indirect indicator of environmental impacts from sulfur emissions, total SOx emissions may be chosen to represent environmental impact I. Then the technology factor T is estimated by dividing I by GDP, and a regression is performed on a quadratic functional form:

$$\ln[T(A)] = a + b\ln(A) + C\ln^2(A)$$

If b is positive and significant, and c is negative and significant, then T must increase at low A and decline at high A – in this case the pollutant in question exhibits the features of an environmental Kuznets curve.

As presented here, in logarithmic form, the regression formula can be used for scenario exercises. In principle it can be applied across a full range of values for A. However, it exhibits a strongly incomedependent elasticity of environmental impact. That is, the per cent change in T with each per cent change in A varies significantly across the full range of incomes. Below, a formulation is proposed that asymptotically approaches a constant-elasticity function at low and high values of A.

An Alternative Form for the Environmental Kuznets Curve

For scenario exercises, where the goal of the modeling exercise is to explore the implications of a scenario narrative in a variety of circumstances, a formulation is needed that will work for a wide range of values of A. Also, it must be relatively simple, with a small number of free parameters. The regression formula above fits these criteria. An alternative form is proposed here, which in addition asymptotically approaches a constant-elasticity function at low and high values of A. The form proposed in this case study is the following one:

$$T(A) = \left(1 + \frac{\varepsilon_{high}}{\varepsilon_{low}}\right) \cdot \frac{\varepsilon_{low} T_{max} \left(\dfrac{A}{A_{max}}\right)^{\varepsilon_{low}}}{\varepsilon_{high} + \varepsilon_{low}\left(\dfrac{A}{A_{max}}\right)^{\varepsilon_{low} + \varepsilon_{high}}}$$

While this may appear complicated, it has a simple behaviour:

☆ When A is at A_{max}, T is equal to T_{max}. At any other value of A, T will be less than T_{max}.

☆ When A is much smaller than A_{max}, T is approximately equal to

$$T(A) \approx \left(1 + \frac{\varepsilon_{low}}{\varepsilon_{high}}\right) \cdot T_{max} \left(\frac{A}{A_{max}}\right)^{\varepsilon_{low}} , \quad A \text{ much less than } A_{max},$$

while when A is much larger than A_{max}, T is approximately equal to

$$T(A) \approx \left(1 + \frac{\varepsilon_{high}}{\varepsilon_{low}}\right) \cdot T_{max} \left(\frac{A}{A_{max}}\right)^{-\varepsilon_{high}} , \quad A \text{ much less than } A_{max},$$

This means that at low incomes, T increases with an income elasticity of ε_{low}, while at high incomes, T decreases with an income elasticity of $-\varepsilon_{high}$. This behaviour has a simple interpretation. As with any elasticity, it means that a 1 per cent increase in income leads to an ε_{low} per cent increase in T at low incomes, and a $-\varepsilon_{high}$ per cent decrease in T at high incomes.

To summarize, with this formulation, there are four parameters, each with a straightforward interpretation:

☆ T_{max}, the maximum value for T.

☆ A_{max}, the income where T reaches T_{max}.

☆ ε_{low} the income elasticity of T at low incomes.

☆ $-\varepsilon_{high}$ the income elasticity of T at high incomes.

The connection between the usual regression curve and the functional form proposed here is as follows. If the full curve behaves like the functional form proposed here, then in the region of the peak it behaves like the regression curve. In that region, the regression curve can be used as an approximation, with

$$a = \ln(T_{max}) + \frac{1}{2}\varepsilon_{low}\varepsilon_{high} \cdot \ln^2 A_{max}$$

$$b = \varepsilon_{low}\varepsilon_{high} \cdot \ln A_{max}$$

$$c = -\frac{1}{2}\varepsilon_{low}\varepsilon_{high}$$

Note that matching the parameters between the regression formula and the formulation proposed here does not uniquely determine the elasticities. Instead, it determines their product. This indeterminacy arises because the regression formula has three parameters, while the formula proposed here has four. This can be seen as an advantage of the formulation proposed in this case study. The regression formula probes only part of the Environmental Kuznets Curve. It cannot be reliably and uniquely extrapolated to values outside of the range where it has been tested. The present formulation is a reminder of this, and allows a scenario analyst to explore alternative assumptions, in that the behaviour of the function at very high and low incomes is not uniquely determined by how the curve looks near the peak.

There is a further theoretical advantage to the formulation offered in this case study. In general, a constant income elasticity function is to be expected when there is no income *scale* that marks a change in economic or environmental regime. In the Environmental Kuznets Curve approach, there is an income scale, set by A_{max}. Very far from A_{max}, however, at high and low income, the economy, and people's behavior should be indifferent to the transition at A_{max}, so it is reasonable that at very high and low incomes the curve should asymptotically approach a constant-elasticity function.

The "Environmental Kuznets Curve": Its Genesis and Hypothesis

For some "limits of growth" advocates such as Georgescu-Roegen (1973) and Meadows *et al.* (1972), the growing economic activity requires more and more energy and material inputs, all the while generating more and more waste by-products. The latter would then undermine the "carry capacity" of the biosphere and result in the degradation of the environmental quality. The degraded stock of natural resources would eventually put economic activity itself at risk and jeopardize the future growth potential. Therefore, to spare both the environment and the economy, economic growth must be halted to make a transition to a more steady state.

In contrast with this radically pessimistic view, the Environmental Kuznets Curve (EKC) assumes that the relationship between various indicators of environmental degradation and per capita income can be depicted thanks to an inverted-U-shaped curve showing that the environment degradation indicators, first boosted by economic growth, shall be decoupled from economic growth trends, and then fall once the income level has reached a given critically high level.

The genesis of the Environmental Kuznets Curve can be traced back to Kuznets (1955), who originally hypothesized that the relationship between inequality in income distribution and income growth follows an inverted-U curve. Since the early 1990s, it has re-gained academic attention. In this line of research, the pathbreaking study was that of Grossman and Krueger (1991) concerning the potential environmental impacts of NAFTA; the Shafik and Bandyopadhyay (1992) provided the background study for the *1992 World Development Report*; Panayotou (1993) wrote part of a study for the International Labour Organization. They all reached the same conclusion: it appeared out of cross-country analyses that the connection between some pollution indicators and income per capita could be described as an inverted-U curve. Panayotou (1993) first coined it the Environmental Kuznets Curve (EKC) given its resemblance to Kuznets' hypothesis.

The EKC Hypothesis and its Various Interpretations Policy-wise

However, from the inverted-U relationship between pollution variation and income growth, economists have derived totally different policy interpretations. At the most optimistic end of the spectrum, Beckerman (1992) argues that the easiest way to obtain environmental improvement is to carry on with the original economic growth path and endure the transient environment deterioration. Claiming both the demand and supply capacity for a better environment increase along with the income, he concludes that, "the strongest correlation between the incomes and the extent to which environmental protection measures are adopted demonstrates that in the long run, the surest way to improve your environment is to become rich". In his very controversial book *The Sceptical Environmentalist*, Lomborg (2001) also describes our future as a "beautiful world". After drawing both environmentally and economically-oriented comparisons regarding various aspects of the world over the last century, he concludes that this world "is basically headed in the right direction and that we can help to steer this development process by insisting on reasonable prioritisation (the economic growth)" (pp. 350-352). Barlett (1994) goes even further in his claims about environment-growth nexus. For him, environmental regulation, as a policy tool reducing economic growth, would actually cause the environmental quality to decrease.

At the other end of the spectrum, pessimists explain that this inverted-U pattern is a consequence of trade liberalization as polluting industries are redistributed among countries of different income levels. "Trade itself is likely to increase the impacts (of pollution) in developing countries and reduce them in the developed countries and this may be another explanation for the EKC relationship" (Suri and Chapman, 1998).

The hypothesis of a "pollution haven" further supports this interpretation: developing countries have comparative advantages in the polluting sectors as their relatively lower income standards cannot cope with such stringent an environmental regulation as their richer trade-partners. Indeed, with trade liberalization, the pollution-intensive industries tend to desert developed countries and move offshore, in developing countries where pollution control is less severe. As mentioned further up, the famous inverted-U curve observed in cross-country experience would then mirror this static transfer mechanism. In this view, the overall worldwide environmental quality may even be more at risk, as developing countries generally have less efficient pollution abatement technologies. Ekins (1997) even points out that if pollution in developed countries is reduced because they discharge their pollution burden, this may not be a potential avenue for today's less developed countries. Therefore, supporting the EKC pattern may impede sustainable development in both the developing countries and the world at large

More economists take a neutral stance concerning the EKC hypothesis: they believe the inverted-U curve only captures the 'net effect' of income upon the environment, in which "income growth is used as an omnibus variable representing a variety of underlying influences, whose separate effects are obscured" (Panayotou, 2003). In this perspective, linking income growth with the easing of environmental deterioration is *not* automatic. To understand the underlying mechanisms that come with economic growth, many authors regard the inverted-U relationship as a "stylised" fact. They strive to "demystify" it, offering and verifying obscured structural explanations for it: better efficiency in production, consumption and pollution abatement; institutional development; improvements in market and institutional efficiency; strengthened public awareness of the negative effects of pollution on health; increasing willingness-to-pay for pollution abatement; and an economic structure that, from one dominated by industrial sectors, becomes dominated by tertiary sectors, etc.[1] From their point of view, economic growth is neither a necessary nor a "sufficient factor to induce environmental improvement in general". (Arrow *et al.*, 1995, p. 92) The same could be said for the turning point between economic growth and environmental degradation yielded by cross-country experience. In some pollution cases, some degree of improvement can be observed even in countries whose income level is still low currently.

Should we Anticipate a Kuznetsian Relationship?

To reiterate, the environmental Kuznets hypothesis predicts environmental damage first increases and then decreases with rising income. Graphically, the relationship should resemble an inverted-U shape. However, referring to the nature of the environmental indicator used in this study, the relationship will be reversed and should resemble a U shape. Otherwise known as a parabola, the regression equation must therefore take the following quadratic form:

$$y = x^2 - x$$

Yet can biodiversity really be expected to exhibit this behaviour? Consider the two halves of our Kuznets curve, the "falling limb" and the "rising limb". The "falling limb" indicates decreasing numbers of species. This is perfectly foreseeable given the well-documented history of extinctions based on habitat loss. We cannot, however, expect biodiversity to be replenished at the same rate. For example, the background species extinction rate has been put at 1 every 100–1000 years (Reid 1994). A creation rate of around the same order of magnitude is likely, so theory would not predict a "rising limb".

Instead, we might expect biodiversity to decrease and then *level off* with increasing income, since economic forces should fuel the drive for environmental improvement but biodiversity cannot replenish itself at an equivalent rate. Thus it would be revealing to estimate a hyperbolic curve as well:

$$y = 1/x$$

Finally, inspection of the data for species diversity also indicates a close clustering of observations around the starting value, independent of income level. In view of this, a linear equation will also be tested:

$$y = -x$$

The conceptual forms of these three curves are presented in Figure 1.

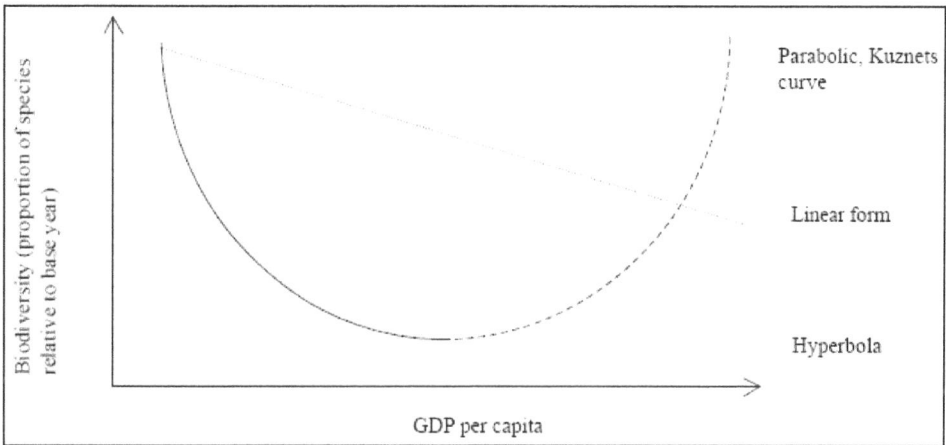

Figure 1: Possible forms of the relationship.

Econometric Considerations

Various techniques exist to estimate the above equations. The two basic methods available are *fixed* and *random effects.* They both build on the regression equation for simple ordinary least squares (OLS):

$$y(i,t) = \alpha + \sum_{j=1}^{\kappa} \beta(j)X(i,t,j) + \varepsilon(i,t)$$

where y is the dependent variable, biodiversity, for country i at time t, a is the equation constant, equal for all countries, b represents the coefficient for independent variable x, where there are j = 1- k variables (each b also being common across all countries) and e(i,t) is a classical disturbance with

$$E[\varepsilon(i,t)] = 0$$

$$Var[\varepsilon(i,t)] = \sigma^2(\varepsilon)$$

Fixed effects loosens up the assumption of commonality across countries by estimating a separate constant a for each country:

$$y(i,t) = \alpha(i) + \sum_{J=1}^{K} \beta(j)X(i,t,j) + \varepsilon(i,t)$$

Random effects works in a similar way but assumes that international heterogeneity is randomly (and normally) distributed. Each country now has a disturbance term m(i):

$$y(i,t) = \alpha + \sum_{J=1}^{K} \beta(j)X(i,t,j) + \varepsilon(i,t) + \mu(i)$$

where,

$$E[\mu(i)] = 0,$$

$$Var[\mu(i,t)] = \sigma^2(\mu)$$

$$Cov[\varepsilon(i,t), \mu(i)] = 0$$

The difference between the two models, whether the vertical displacement of the regression equation should be parametric or random – has been debated in the literature (Greene 1997). Fixed effects imply international differences are generated by country-specific factors not covered by our regressors. Random effects, on the other hand, imply national peculiarities are unimportant and differences should be assumed random. This is a responsible approach if the sample to be studied is part of a much larger population (Greene 1997). Yet here we have an almost complete set of countries containing significant tropical moist forests. Furthermore, intuition suggests environmental and economic factors should play a rather important role. We can test this empirically using the Hausman test but, at this interjection, we would expect fixed effects to be favoured.

Koop and Tole (1999), however, charge both of these specifications with assuming too much commonality and recommend the additional freedom of the *random coefficients* method. In this case, each country possesses its own curve, drawn from a random distribution:

$$y(i,t) = \sum_{J=1}^{K} \beta(i,j)X(i,t,j) + \varepsilon(i,t)$$

where,

$$\beta(i) = \beta + v(i)$$

$$E[v(i)] = 0$$

Thus we are no longer bound to the assumption that each country's curve possesses the same shape, yet we are not going so far as to say each country has its own curve which is unrelated to all others (this is the basis of the *fixed coefficients* model).

However, the nature of the data in this study is such that the test for random coefficients is distorted. Consequently we shall confine ourselves to comparing fixed and random effects empirically.

To summarise, the variables and equations are as follows:

S(i,t) is the fraction of species in any year compared to the reference year 1970 (equation 2). This tends to cause the data to cluster around the reference mark 1.0, so all values are multiplied by a factor of 1000.

G(i,t) is income per capita. It is always converted to log form. Data comes from the Penn World Table.

P(i,t) is population change, expressed as a percentage of the previous year. Data comes from the Penn World Table.

D(i,t) is population density, expressed as people per hectare. The variable is a combination of FAO Production Yearbook data on land area and Penn World Table population statistics.

T(t) is a linear time trend (T is simply the relevant year) used to correct the problem that both biodiversity and income display positive time dependency.

F(i,t) is forest area in hectares. The relative impact of deforestation in any country depends on absolute forest area and feeds back into future trends.

e(i,t) is a classical error term.

The basic regression model for the quadratic equation is

$$S(i,t) = \alpha + \beta_1 \ln G(i,t) + \beta_2 [\ln G(i,t)]^2 + \beta_3 P(i,t) + \beta_4 D(i,t) + \beta_5 T(t) + \beta_6 F(i,t) + \varepsilon(i,t)$$

For the hyperbolic equation, it is.

Chapter 6
Environmental Valuation

Environmental deterioration can be defined as "the loss of capital nature". "Environment accounting" becomes the paradigm of the conservation and preservation of such capital by the same standards of an enterprise patrimony. Environmental goods valuation can be a prerequisite in order to control and contain the damages caused by man to the environment.

From the cowboy economy attitude, according to which the natural environment had to be conquered and civilized in conformity with the idea of the open system and that of continuing economic growth (abundance of resources, expensive use of energy),we have passed to a different perception of the environmental problem, the spaceman economy. The Earth-Spaceship metaphorie the consideration of a circular economic system, has given prominence to the typical limited aspect of environmental resources. Environmental deterioration main artificer is the industrial and commercial "criminal development", permitted by the incessant technical and scientific acceleration. Nevertheless the deterioration accomplice is often "the missing awareness and determination of the total economic value of resource and natural functions" (Barbier 1989). The environmental conceptual apparatus is that of the Economy of Wealth. In contrast with some of these theoretic presuppositions, a second discipline has been delineated: the ecological economy, whose epistemological principles are different. The ideological visions concerning environmental problems can be summarily assembled into two general categories: "technocentrism" and "ecocentrism". The former category includes the positions considering, on different levels, the natural patrimony whose aim is the satisfaction of human needs; the category of ecocentrism enlarges the ethic reasoning and arrives at ascribing rights, moral interests to non-human species, even to environmental a biotic components. The environmental patrimony protection is by now a prerogative of developed countries. But its effective accomplishment is hampered by a conjunction of factors deriving from the difficulties in valuating the shocks undergone by the system; from the rapid and sufficient realization of their presence, so to prearrange appropriate countermeasures; from the achievement of a difficult balance between misuses containment and pollution from one side, and stimulation of the industrial production from the other. According to the contemporary economic perspective, environmental goods such as air, water, fauna are valuable goods, since the offer a flow of services to the individuals. In the service economic value the measuring process of services, supplied by natural resources, is a part of the benefit/cost approach. In conformity with this statement an enlarged point of view should be adopted, so to make the services real flow supplied to society and economy in natural resource readable: before all as an input source (fossil fuels, lumber, minerals, etc.), secondly as an indispensable element for human life (breathable air, livable climatic condition, etc.), then as a supplier of a series of reconstructive and landscape opportunities and finally as a system capable to receive and waste the surplus coming from the human activity. Consequently, the environmental economic value can be defined as the sum of flows discounted net values deriving from all the offered services: the benefits of an increasing support of any environmental service flow are given by the increase of the service discounted value. Likewise pollution damages correspond to the reduction in the service flow. The used value concept is founded on the economy of wealth: the individual wealth/utility dose not depend only on consumed – public or private – but even on the quantity and the quality of nonmarket goods and services supplied by natural/environmental resources system (*e.g.*, health, recreative opportunities, landscape services, etc.). It follows that the reference for the economic value derivation measures of changements in the resources/environment system is the effect on human wealth.

If society wishes exploiting the equipment of natural resources in the most efficient way, the values of goods/services flows coming from the resources use itself (*i.e.*, the benefits)and enjoyed by every member, should be compared to the values they renounce deterring environmental goods/services from other employments (*i.e.*, the costs). Since the benefits and the costs are valuated according to their effect on the individual wealth, the "economic value" and "economic value" concepts correspond. The economic theory maintains that the individuals have proper preferences among goods/services alternative bundles – of market or not and that preferences enjoy the replace ability property among such goods/services. Some observers are critical towards the attempt of expanding the economic measure process to elements such as health, human security, environmental features, landscape values and synthesizing its value in a monetary measure. Substantially the economic approach to the environment, whose quantification is presumed and its specify dispersed in the homogeneous measure of currency, has been under discussion.

Genesis of the Total Economic Value

Environmental values are supported and reinforced in their informative content by the economic evaluation in confront of other values present in decisional process. In this condition of inferiority it risks to take the decision of altering the environmental resource irreversibly. If the development benefits can be easily monetized, those of preservation/conservation are not the same because it needs to consider interests and aims of the individuals, who are not necessarily direct consumers of goods. An unbalanced exploitation directed to market values runs the risk to put in danger the considered environmental goods in a medium-long period.

If risen to a unique evaluative criterion, exchangeable values can irreversibly compromise the elements forming the social cultural value of a resource and schematically connectable to its quality. The Total Economic Value has made the idea of plausibility natural: direct use values refer to the economic dimension, those of indirect use refer to the ecological dimension. The vicarious and legacy values are linked to the social dimension. As the sustainable development aims at reaching a compromise among its three distinguishing dimensions, the total economic value should consider the trade-off of its parts. For example the total economic value of a wood cannot include factory lumber and that for burning obtained by a smooth cutting of existing trees, otherwise other use values, even non-use values, would be excluded, so that the total value would exclusively correspond to the cut lumber use value; on the contrary following and occasional cuttings are compatible with other expressions forming the total economic value. The total economic value of an environmental resource may assume two connotations: if sustainable use benefits are prevailing there will be a total value concerning the development; if non-use benefits are favorite there will be a total value concerning the preservation. Between the two shades the prevailing one depends on various factors. The principle one are the features of unreproduceability, rareness, singleness typical of the considered resource, the purer or pureless public goods, the location of the resource, its property rights, the diffusion level of the sensibility on environmental themes.

Criticism to Environmental Goods Monetary Evaluation

It refers essentially to three problems:

a. The problem of knowledge
b. The problem of incongruity
c. The problem of composition

The first problem is linked to what has previously been said about primary and secondary values. According to the authors the "functional transparency" is the range of services supplied by an ecosystem, and the expression underlines the difficulty of identifying and evaluating some eco-systemic functions which remain transparent and unknowable until when they paradoxically stop to be active and environmental damages are so outlined. In presence of environmental goods whose functional transparency is not a basic element, the evaluation task would be easier, but it runs into the comparison among different scales, always concerning the problem of knowledge. In other words it deals with the difficulty of translating into monetary terms the environmental goods attributes running the risk that the preferences concerning goods are influenced by the format and context survey (disputes on contingency evaluation mistakes).

The problem of incongruity refers to a situation in which there is a trade-off between the economic calculation and the oral aspect of the choice. For example essays by many authors underline how in front of certain environmental problem the individuals act more like "citizens" than like "consumers". Statements like "all wild species have the right to live apart from every benefit or damage for people" registered during application of a contingent valutative method, expresses the rejection and the unavailability of people to consider market transactions as choices concerning resources seen as public. Two dimensions coexist in the value attributed to a resource: the former represents the individual utility, the latter the social utility which reflects altruistic motivations concealed under a lot of choices. In the traditional psychology, a similar idea is "the pyramid of needs." According to this interpretation of the human behavior, at the basis of the pyramidal structure there are material needs, first of all the psychological ones (food home etc.) and those of security.

Once satisfied the first need, the social needs arise the sense of belonging, love, selffulfillment, being esteemed. At the hierarchical top there are moral needs like justice. The satisfaction of the need that occupies a higher hierarchical position derives from a selffulfillment process and a development of individual potentialities.

Self-fulfilled people could be so pressed for the behavior through motivations of high environmental responsibility and show value expressions considering the non-use aspect of the resource.

The third problems concerns the total economic value directly, and exactly its forming element. The distinction between use-value and non-use value implies a kind of "reduction to items" of goods, environmental services, though they do not exist in moderate units like any market goods.

So it would be necessary a holistic approach that gives prominence to the uniqueness or not of the environmental goods to evaluate not in relationship with human beings, but with the whole system of which they are part.

Environmental Values and Environmental Economic Values

In economy, the expression "environmental values" means essentially two relationships, which are not at the basis of the decisions concerning the environmental politics broadly speaking. None of these two meanings has to be confused with that of the "economic value".

Inside the category of environmental values, expressed through preferences, a first distinction is done between held values and assigned values (Boulding, 1966). The first ones are the values that influence constantly choices and individual actions and represent advisable behaviour. The second ones express the relative importance of an object in a certain context for an individual or for a group. The assigned value is not a feature of the object but it represents the situation in which the object is in front of other objects; it is the concept of the relative value, not absolute as the held value. The ways in which held values and assigned values express themselves are different; they depend on the presence or not of individual or collective environmental values. Considering the private individual preferences, the assigned values are expressed in terms of willingness to accept and in presence of market failures, it is possible to remedy thanks to surveys and questionnaires (contingent valuation method) and proxy variable price (travel cost method, hedonimetric prices). On the contrary, with collective preferences, the held values should influence individual preferences and form norms operativable through laws and regulations.

An assigned value, concerning the individual preferences, is the total economic value, while we approach to collective preferences in terms of fixed standards.

Various factors concur to form the valuating process: the existing information in the previous moment than that in which we are pushed to decide about influences either held values, with that complex group of inclinations generally allotted and relatively stable in time, or the knowledge itself of the object (direct or indirect).

At the moment of the valuation, new information can be given about the object and this interacts with relevant convictions (*e.g.,* the importance of biodiversity) and perceptions concerning the object to evaluate (*e.g.,* the continuing decrease of a wild specimen). Consequently, according to the individual value perspective it emerges the importance of the motivations concealed under the unobservable sensation of value (utility) that takes the form of assigned value. The role of the pre-existing or new information and of the reasons concerning the use and the nonuse are of great interest because of their significance in explaining how an individual assigns a value to a certain object.

For example, in the surveys that point out the willingness to pay or the willingness to accept, the differences between a measurement and another one can be explained by the motivations involved in the decision, in the same way the result of the valuation is influenced by the information supplied during the survey.

Aims of the Environmental Goods Economic Evaluation

Until the 60s and the 70s natural goods economic valuation had been a typical American practice; in the following years and till today it has assumed an increasing importance in Asian, Latin American and African countries and, in minor measure, even in Europe. As witness of such a diffusion, it can be considered the proliferation of guides to evaluate natural resources written by the main international organizations (UNEP, WHO, World Bank), concerning most of all developing countries and plenty of manuals edited for U.S.A.

The economic analysis of the natural environment can be employed according two main ways inside evalutative processes of environmental effects. Such ways are:

Calibration of public works, analysis of ante-post of resource natural damages, and public works: the non-market goods evaluation, such as the environmental ones bears as a part of public works evalutative process in the first half of the 20th century in the United States aiming at incorporating systematically the intangibles in the economic analysis.

The monetary values attributed to the benefits deriving from the natural resource exploitation through the definition of economic methods, occurred between the 50s and the 60s, represent an important further qualitative leap.

From the 60s, the economic analysis has become the usual instrument supporting public plans (from the hydraulic ones to transportation, health and education/ formation). In Europe the identification of the potentialities, both theoretic and applicatory, of the economic analysis is more recent in confront of what has happened in the U.S.A. and the methods development has proceeded more slowly. In certain countries the economic analysis has been used as a supporting instrument for public choice mostly for what concerns road infrastructure, but environmental impacts have not usually been considered.

Environmental Resource Damages

The environmental externalities quantification is relevant to valuate natural resource damages. To this purpose, in U.S.A. the most controversial and deepest inter-relationship between environmental evaluation and public choice occurred when the Congress promulgated the Comprehensive Environmental Response Compensation and Liability Act in 1980. It foresaw the creation of a super-fund to finance the drainage of the existing dumps of dangerous waste materials and established the responsibility of the involved parts to indemnify the damages caused to natural resource because of these sites.

The regulamentations aims at allowing the damaged resource return to the previous condition and compensating the involved parts for each natural resource service loss through the recovery, rehabilitation, replacement or the acquisition of equivalent services. The regulamentations define the evalutative process in three phases:

☆ Pre-evaluation: to establish how the recovery can be achieved;

☆ Planning: to identify the needs and the aims of recovery activities;

☆ Accomplishment: to make the planned activities effective.

In Europe the concept of responsibility in the evalutative processes of "natural resource damages, conducted through non-market evalutative methods have not been considered by the same dignity until yet.

The Total Economic Value

The expression of total economic value bears as an attempt to overcome the traditional evaluation of environmental goods, exclusively based on the use value attributed to goods considering direct benefits enjoyed by final consumers. It seems that the expression "total economic value" appeared for the first time in an essay by Peterson and Sorg in 1987, "Toward the measurement of total economic value". Then the term was more and more used by other environmental economists, among whom Turner and Pearce (1996). The use value derives from a concrete use of environmental goods. Even the value attributed to goods to individuals is included in the use value, because they enjoy to see a landscape or the can swim in a lake; even those ones can be considered users of environmental goods, even if in a unappropriate and under-destructive manner. Every use, in any moment and by anyone are realize to create use values, which are more or less measurable since they derive from their current use.

But the total economic value is not only use value; it is given by the sum of use and nonuse values referring to intrinsic benefits, *i.e.*, those deriving from the mere existence of environmental goods. The first economist, who identified the total economic value double feature, was Kutrilla (1995). After Kutrilla the scholars interested in this topics have not been limited to theorical analysis of the total economic value and of its components, but their attention is centred on an empirical analysis which allows them to identify the main features especially of non-use value and the different methods usable for their measurement.

The uniqueness and irreversibility play a central role considered that, according to these features, certain individuals, even uncommon users of goods express the willingness to pay a tax so to allow goods to remain in such a way (in the case a park is being closed). The particularly innovative element is the explicit reference to economic subjects who, without using the goods, can be interested in its conservation. In this context it is the first time that the so-called "option value" has been delineated, *i.e.* the maximum amount that the non-users are willing to pay so that the park can stay open. In particular a central role about these subject matters was played by John Kutrilla. At the end of the 60s he conducted an analysis in which in contrast with the use value, he identifies a larger concept of non-use value, which includes, besides the option value, other two components: the existence value and bequest value. The existence value is defined as "the value attributed to environmental goods by the economic subject without a link to a real or potential use, but exclusively to their mere existence". An economic subject is willing to pay a certain sum in order to avoid the destruction of any environmental goods, such as a park or a forest, that is not why he intends to visit such goods in the future, but he simply wants the goods continue to exist. The other component of nonuse value is the "bequest value", defined as the value that an individual attributes to goods considering the use of the goods in the future by his heirs". The bequest value originates from the third motivation identified by Krutilla, *i.e.* the one that is linked to the individual's willingness to pay so that certain goods can be conserved for the posterity. Consequently such value becomes a non-use value if we refer to the contemporary generation, which is not interested in the fruition of goods and results to be a potential use value for future generations.

Direct and Indirect Use Values

Direct use values are produced in consequence of an immediate or mediate contact with the resource, the environmental goods. The access and the use are of two levels: the primary one, where the physical and immediate contiguity to the resource is a necessary condition to gain benefits, and the secondary one, where the fruiter has not a direct relationship with the resource.

Consumption use value refers to "extractive" activities, whose object is a precise resource "consumable" in the primary manner (*e.g.*, through hunting, picking and gathering wild fruits) or in the secondary manner entering other goods (natural substances present in some medicines; the ivory of elephants' tusks) as a productive factor.

Non-consumption use values refer to all those activities that exploit the resource for recreative and amusing purposes, without its material consumption. A walking tour in the mountains and the bird watching are some examples of these activities which do not cause any damage to the resource, obviously excluding episodes of congestion; on site research activities are considered among non-consumption primary values.

The vicarious use value refers to an off-site resource fruition, an example can be given by the reading of an article about a dying species or watching a documentary on nature.

Some research values are included among secondary values. In this category there is the information obtained by certain studies on animals or vegetables, as the research on birds in order to survey the store of pesticides in the environment and the use of some vegetables to survey the atmospheric pollution (*e.g.*, the use of tobacco plant as ozone bioindicator or forage grass for heavy metals). Even if the research is often conducted necessarily on the spot, they are considered secondary values, because the obtained data are employed in places farer than those where the resource is normally present.

Indirect use values refer to regulating ecological functions carried out by the system and converged in the general categories of functions supporting life and the pollution control. The indirect use comes from the implicit carried out in supporting or protecting economic activities. For example, accumulation functions of underground and artificially recharged water in some damp areas (flooded plays and beat bogs) are used indirectly, because water is used for domestic and agricultural purposes.

The valuation techniques employed in the indirect values estimation are based essentially on market values given by environmental defensive expenses subdivided into three categories:

1. Preventive expenditures for the environment;
2. Avoidance costs;
3. Treatment of damages;

Preventive expenditures refer to sustained costs in order to avoid a environmental negative external effects (*e.g.*, the introduction of "clean" techniques and processes, the depurators), repairing expenditures refer to restoration costs of damaged environmental functions (*e.g.*, polluted reclaimed lands).

Avoidance costs are those supported to treat negative external effects (*e.g.*, the installation of sound absorbent barriers along the roads which are near the towns to protect from acoustic pollution).

Treatment of damages refers to costs supported to compensate individuals or goods for suffered damages (for example, in the first case medical expenses concerning diseases due to environmental deterioration; in the second case restoration and cleansing expenses of the monuments damaged by urban pollution).

Difference between Use Value and Non-Use Value

The concept of Total economic value requires a precise distinction between use value and that of non-use. Knowing one of these components it is possible to obtain the other subtracting it from the total value. To illustrate this pattern, two groups of sets are introduced, considered the basic levels of the resource and its costs. In the first set the resource is under its basic level, while the costs are settled. Y_1 is defined as minimum income required by an individual to sustain the utility at its basic level. In the second set, costs and resource are ah their basic levels; Y_2 is the minimum amount of the income to sustain the utility firm in its level

The resource total value can be defined by the difference between the minimum income of the first set and that of the second one:

$$T = Y_1 - Y_2$$

T represents the minimum income which makes an individual indifferent between a set whose resource is at its usual level and another one whose resource is lowered qualitatively and quantitatively.

A second group of sets is employed to reveal non-use values.

In the third set the resourced is more reduced than its current level and its complementary goods have high costs so to choke prices. In this set the individual is not a non-user of the resource and Y_3 is the minimum income required to keep utility at the basic level of the other sets.

In the fourth set, the costs of complementary products are still choke prices, but the resource is at its basic level; Y_4 is the income that sustains the utility constantly. Since in these sets the individual dose not use the resource, the difference between Y_3 and Y_4 represents the component of non-use values.

$$NUV = Y_3 - Y_4$$

So the use value is obtainable:

$$UV = T - NUV$$

Being available an estimation of use values, these ones can be deducted from the total value obtaining non-use values.

Incidental Value and Vicarious Consumption Value

Incidental use means a form of utilization of a resource, mostly referred to nonconsumption, that an individual can experience a very occasional way without the necessity to buy additional goods. You may think of an individual who lives in the area of a natural parks and sees a deep from the window of his house or going to work. Certainly hedonimetric techniques can be used, as in the second case, or we can consider the time value employed in these occasional uses, but it remains the eventuality that not all incidental uses leave marks in the tendency of the market and in the time allocation. Other methodological unknown factors appear with the vicarious use value, which includes the purchase of books, magazines, or the vision of documentaries about a particular environmental resource. The main one is represented by the fact that out of the resource to evaluate these additional goods often contain information concerning other environmental goods and it is difficult, even impossible, to assign a value quota to the specific resource (Freeman, 1984).

The vicarious use value can be distinguished on the basis of the features of media used to create It. If an individuals enjoys the resource through pictures (or taped video-cassettes) taken by himself, the use value can be analysed in the relationships between the resource and the input request of photos and videos production. On the contrary, if the resource is used through the vision of T.V.

programmes or the reading of magazines, the relationship between the resource and the information is complex: it can occur that the information request increases as a consequence of an environmental disaster so to stay in the increasing vicarious use value paradox corresponding to a damage caused to the resource.

Price Option and Option Value

The price option (PO) is defined as the maximum amount that an individual, without certain preferences, is disposal to pay to gain the option to visit the park in the future. Once solved the doubt ad establishing he is one of its fruiters, the consumer surplus is the sum the individual accepts to pay to visit the park. The expected consumer surplus E(CS) is the result of the relationship between CS and the probability of the will to visit the park.

The option value is the result of the difference between price option and expected consumer surplus

$$OP=PO-E(CS)$$

Relying exclusively on E(CS) within the decision to jeep open or close the park means undervaluing the resultant benefits of the choice to keep it open. But this deduction is based on two presupposition: according to the first the price option is the right measure to use with these two kinds of decisions; according to the second the option value is positive for those who are unfavourable to any kind of risk.

Quasi Option Value

This second interpretation of option value, called quasi option value (QOV), is centered on inter-temporal aspects of uncertainty giving prominence to the role of irreversible decisions and of information flow available in the time.

The quasi option value represents the benefit connected with the postponement of the decision about the resource irreversible development in presence of the doubts on the benefits deriving from its preservation.

The same conclusion, drawn about the option value, regards the inclusion of this kind of value in the total economic value; in literature according to the prevailing opinion it does not represent a benefit distinct and separated component, but it is shaped as an information value.

Non-Use Values and Use Value

Non-use values are independent from any benefit linked to the use of an environmental goods; these values are connected with the prolonged existence of goods, without any kind of contemporary or planned use.

The non-utility, the sorrow felt by a lot of people in learning that a seciesmen is dying out or that a wood has been destroyed by a fire, the are some examples which witness the presence inside the individuals of this category of values, even if the will not visit the wood threatened by the fire or the dying animal.

Consequently the resource continuous existence is a prerequisite of non-use values, so that in literature the expressions non-use value and existence value are employed interchangeably, as synonyms (Signorello, 1992) or meanings considering the first ones given by the option value and the existence value, generally the last one meant as comprehensive of all values according to the condition of certainty referring motivationd different from those referred to personal use (Bishop and Woodward, 1995).

Other expressions that may be found in literature referring to non-use values are:

☆ the intrinsic value given by the sum of option, aesthetic, legacy and existence values;

☆ the preservation value made of option, legacy and existence values;

☆ he intangibles made of existence, legacy, option values and of vicarious use;

☆ non-user values;

☆ passive use values meaning the absence of a behavioral evidence;

☆ off-site use values (Randall and Stoll,1980).

The meanings characterizing the use value of an environmental resource are several, The narrow vision of NUV separates these ones from any relationship with other goods, so that use values are characterized by any complementarity between the resource and market goods (enlarged vision of UV). An alternative method to make a distinction between UV and NUV is not based on the identification of activities and motivations, which can mark them, but on the verification of the individual condition of the resource user and non-user (Sellin and Rosato,1998).

From the practical point of view the supposition,that the visitors have only use values and the non-visitors the non-use ones, avoids the problems concerning the difference between use and non-use. Such planning is weak on the theoric level, because the logic does not exclude that even the users refer to non-use values. A close observation of the possible definitions concerning the non-use notions is inconclusively; by definition, if the use values is linked to the on site use of the resource (through the purchase of some additional goods), this definition has the merit to distinguish among situations in which the use of the resource is measurable through the methods of the travel cost and of the hedonimetric price and among situations in which it is not possible because of the presence of the vicarious or incidental values.

Existence Value

There are two basic definitions of the existence value (EV) found in literature and, if no substantial difference can be notice between them, a closer observation reveals some meaning shades that develop on the methodological level. A first formulation of the EV gross up to Kutrilla (1985) (long version), who maintained that environmental resource (specifically the wilderness) can have a value also for those people who gain satisfaction and pleasure by the mere knowledge of the existence of that goods (in a continuous manner).

A second expression of the EV (short version) refers to the willingness to pay for preservation, protection and qualitative and quantitative increasing of natural goods. According to this definition, the information about the resource and that about its existence are interrelated and this relationship explains how the vicarious use value, which is always a use value, is difficult to distinguish from the existence value (Randall and Stoll, 1980; Pearce, 1989).

The EV itself, appealing to the availability to pay, includes motivations referable to the non-use notion due to altruistic attitudes.

Randall and Stoll makes a list of three kinds of altruism:

☆ intragenerational altruism;

☆ intergenerational altruism;

☆ Q-altruism.

The first type of altruism, defined also philanthropic altruism or vicarious value (VV), refers to a resource evaluation not based on personal use considerations, but on the opportunities that other people, contemporary to the examiner, are able to use the considered goods. Through the vicarious value, the altruist can verify if the motivations revealed by the beneficiary about the resource are of use or non-use.

The intergenerational altruism, which is at the base of the legacy value (LV) or of the bequest value, moves from the idea that the contemporary generation, motivated by this kind of altruism, wishes to transmit the most possible undamaged equipment of contemporary resource to future generations. According to some scholars, the legacy value is seen as a nonuse value, while according to others it is connected with the option use value and with that of non-use.

This classifying variety that may be found in literature is another confirmation of the missing achievement of a theoric agreement, and consequently terminological, about the total economic value concept, which is still susceptible of new defining contributions.

The meanings of the altruism, which are under the vicarious value and legacy value, are "domestic" altruism and "diffuse" altruism so to include all the people in general. The altruism of the scenery of the legacy and vicarious value is called "paternalistic". The utility function of the altruist includes, besides the goods consumed directly by the individual, also the environmental goods object of the altruistic motivation.

The third and fast kind of altruism introduced by Randall and Stoll (1980) is the Qaltruism, also called intrinsic altruism, and it is based on the knowledge that the Q-resource itself benefits by remaining undisturbed longer than possible and integral in its functions. If in the other two cases the beneficiaries of the altruistic act are contemporary or future people, on the contrary with the intrinsic altruism the beneficiary is the resource itself, and the role played by the human being is that of giving voice to this intrinsic right of existence, the possessory title of the resource. The analytic treatment of the altruism in favor of a resource is difficult. Before all it is difficult to reprent economically the environmental goods wealth, and we can add to it the marginal character assigned to the individual in this perspective, that of the spokeperson of the resource interests. In the consideration about the motivations connected with the existence value, the intrinsic value pays particular and special attention to life conditions of non-human species and to the health conditions of whole ecosystems. The Q-altruism is similar to the concepts of sympathy for other living beings and of steardship.

Market Price Method

The market price method was selected in this case, because the primary resource affected is fish that are commercially harvested, and thus market data are available.

Application of the Market Price Method

The objective is to measure total economic surplus for the increased fish harvest that would occur if the pollution is cleaned up. This is the sum of consumer surplus plus producer surplus. Remember that consumer surplus is measured by the maximum amount that people are willing to pay for a good, minus what they actually pay. Similarly, producer surplus is measured by the difference between the total revenues earned from a good, and the total variable costs of producing it. Thus, the researcher must estimate the difference between economic surplus before the closure and economic surplus after the closure.

Step 1

The first step is to use market data to estimate the market demand function and consumer surplus for the fish before the closure. To simplify the example, assume a linear demand function, where the initial market price is $5 per pound, and the maximum willingness to pay is $10 per pound. The figure shows the area that the researcher wants to estimate the consumer surplus, or economic benefit to consumers, before the area was closed.

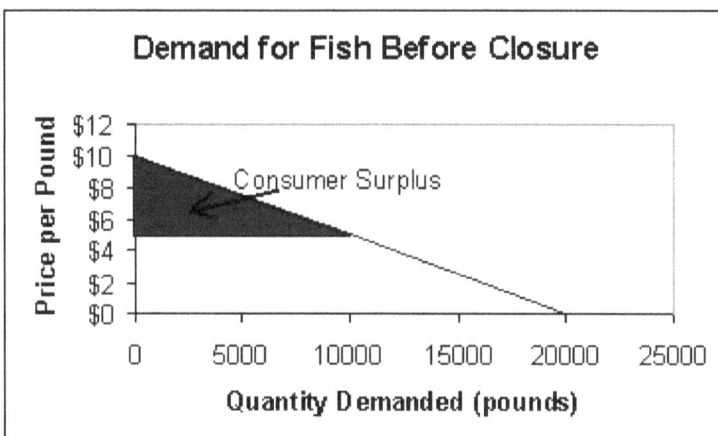

Figure 1

At $5 per pound, consumers purchased 10,000 pounds of fish per year. Thus, consumers spent a total of $50,000 on fish per year. However, some consumers were willing to pay more than $5.00 per pound and thus received a net economic benefit from purchasing the fish. This is shown by the shaded area on the graph, the consumer surplus. This area is calculated as ($10-$5)*10,000/2 = $25,000. This is the total consumer surplus received from the fish before the closure.

Step 2

The second step is to estimate the market demand function and consumer surplus for the fish after the closure. After the closure, the market price of fish rose from $5 to $7 per pound, and the total quantity demanded decreased to 6,000 pounds per year.

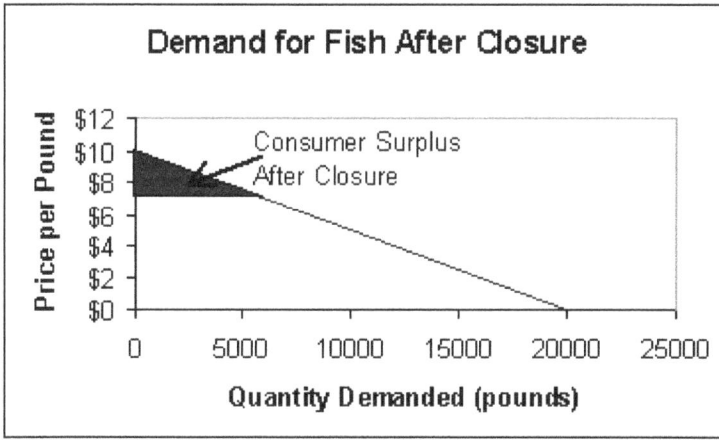

Figure 2

Thus, the economic benefit has decreased, as shown in the figure. The new consumer surplus is calculated as ($10-$7)*6,000/2 = $9,000.

Step 3

The third step is to estimate the loss in economic benefits to consumers, by subtracting benefits after the closure, $9,000, from benefits before the closure, $25,000. Thus, the loss in benefits to consumers is $16,000.

Step 4

Because this is a marketed good, the researcher must also consider the losses to producers, in this case the commercial fishermen. This is measured by the loss in producer surplus. As with consumer surplus, the researcher must measure the producer surplus before and after the closure and calculate the difference. Thus, the next step is to estimate the producer surplus before the closure.

Producer surplus is measured by the difference between the total revenues earned from a good, and the total variable costs of producing it. Before the closure, 10,000 pounds of fish were caught per year. Fishermen were paid $1 per pound, so their total revenues were $10,000 per year. The variable cost to harvest the fish was $.50 per pound, so total variable cost was $5,000 per year. Thus, the producer surplus before the closure was $10,000–$5,000 = $5,000.

Step 5

Next, the researcher would measure the producer surplus after the closure. After the closure, 6,000 pounds were harvested per year. If the wholesale price remained at $1, the total revenues after the closure would be $6,000 per year. If the variable cost increased to $.60, because boats had to travel farther to fish, the total variable cost after the closure was $3,600. Thus, the producer surplus after the closure is $6,000–$3,600 = $2,400.

Step 6

The next step is to calculate the loss in producer surplus due to the closure. This is equal to $5,000–$2,400 = $2,600. Note that this example is based on assumptions that greatly simplify the analysis, for the sake of clarity. Certain factors might make the analysis more complicated. For example, some fishermen might switch to another fishery after the closure, and thus losses would be lower.

Step 7

The final step is to calculate the total economic losses due to the closure – the sum of lost consumer surplus and lost producer surplus. The total loss is $16,000 + $2,600 = $18,600. Thus, the benefits of cleaning up pollution in order to reopen the area are equal to $18,600.

How Can the Results be Used?

The results of the analysis can be used to compare the benefits of actions that would allow the area to be reopened, to the costs of such actions.

Applying the Market Price Method

The market price method uses prevailing prices for goods and services traded in markets, such as timber or fish sold commercially. Market price represents the value of an additional unit of that good or service, assuming the good is sold through a perfectly competitive market (that is, a market where there is full information, identical products being sold and no taxes or subsidies).

Application of the market price method requires data to estimate consumer surplus and producer surplus. To estimate consumer surplus, the demand function must be estimated. This requires time series data on the quantity demanded at different prices, plus data on other factors that might affect demand, such as income or other demographic data. To estimate producer surplus, data on variable costs of production and revenues received from the good are required.

Advantages of the Market Price Method

☆ The market price method reflects an individual's willingness to pay for costs and benefits of goods that are bought and sold in markets, such as fish, timber, or fuel wood. Thus, people's values are likely to be well-defined.

☆ Price, quantity and cost data are relatively easy to obtain for established markets.

☆ The method uses observed data of actual consumer preferences.

☆ The method uses standard, accepted economic techniques.

Issues and Limitations of the Market Price Method

☆ Market data may only be available for a limited number of goods and services provided by an ecological resource and may not reflect the value of all productive uses of a resource.

☆ The true economic value of goods or services may not be fully reflected in market transactions, due to market imperfections and/or policy failures.

☆ Seasonal variations and other effects on price must be considered.

☆ The method cannot be easily used to measure the value of larger scale changes that are likely to affect the supply of or demand for a good or service.

☆ Usually, the market price method does not deduct the market value of other resources used to bring ecosystem products to market, and thus may overstate benefits.

The contingent valuation method (CVM) is used to estimate economic values for all kinds of ecosystem and environmental services. It can be used to estimate both use and non use values, and it is the most widely used method for estimating non-use values. It is also the most controversial of the non-market valuation methods.

The contingent valuation method involves directly asking people, in a survey, how much they would be willing to pay for specific environmental services. In some cases, people are asked for the amount of compensation they would be willing to accept to give up specific environmental services. It is called "contingent" valuation, because people are asked to state their willingness to pay, *contingent* on a specific hypothetical scenario and description of the environmental service.

The contingent valuation method is referred to as a "stated preference" method, because it asks people to directly state their values, rather than inferring values from actual choices, as the "revealed preference" methods do. The fact that CV is based on what people say they would do, as opposed to what people are observed to do, is the source of its greatest strengths and its greatest weaknesses.

Contingent valuation is one of the only ways to assign dollar values to non-use values of the environment–values that do not involve market purchases and may not involve direct participation. These values are sometimes referred to as "passive use" values. They include everything from the basic life support functions associated with ecosystem health or biodiversity, to the enjoyment of a scenic vista or a wilderness experience, to appreciating the option to fish or bird watch in the future, or the right to bequest those options to your grandchildren. It also includes the value people place on simply knowing that giant pandas or whales exist.

It is clear that people are willing to pay for non-use, or passive use, environmental benefits. However, these benefits are likely to be implicitly treated as zero unless their dollar value is somehow estimated. So, how much are they worth? Since people do not reveal their willingness to pay for them through their purchases or by their behavior, the only option for estimating a value is by asking them questions.

However, the fact that the contingent valuation method is based on asking people questions, as opposed to observing their actual behavior, is the source of enormous controversy. The conceptual, empirical, and practical problems associated with developing dollar estimates of economic value on the basis of how people respond to hypothetical questions about hypothetical market situations are debated constantly in the economics literature. CV researchers are attempting to address these problems, but they are far from finished. Meanwhile, many economists, as well as many psychologists and sociologists, for many different reasons, do not believe the dollar estimates that result from CV are valid. More importantly, many jurists and policy-makers will not accept the results of CV. Because of its controversial nature, users must be extremely cautious about spending money on CV studies and about using the results of CV studies.

This section continues with some example applications of the contingent valuation method, followed by a more complete technical description of the method and its advantages and limitations.

Hypothetical Scenario

A remote site on public land provides important habitat for several species of wildlife. The management agency in charge of the area must decide whether to issue a lease for mining at the site. Thus, they must weigh the value of the mining lease against the wildlife habitat benefits that may be lost if the site is developed. Because the area is remote, few people actually visit it, or view the animals that rely on it for habitat. Therefore, non-use values are the largest component of the value for preserving the site.

Why Use the Contingent Valuation Method?

The contingent valuation method was selected in this case because of the importance of non-use values, and their potentially significant levels.

Alternative Approaches

Since non-use values are significant, and few people actually visit the site, other methods, such as the travel cost method, will underestimate the benefits of preserving the site. In this case, contingent choice methods might also be used, depending on the questions that must be answered, and whether contingent choice question formats are more effective than standard contingent valuation questions. This would be decided in the survey development stage of the application.

Application of the Contingent Valuation Method

Step 1

The first step is to define the valuation problem. This would include determining exactly what services are being valued, and who the relevant population is. In this case, the resource to be valued is a specific site and the services it provides – primarily wildlife habitat. Because it is federally owned public land, the relevant population would be all citizens of the U.S.

Step 2

The second step is to make preliminary decisions about the survey itself, including whether it will be conducted by mail, phone or in person, how large the sample size will be, who will be surveyed, and other related questions. The answers will depend, among other things, on the importance of the valuation issue, the complexity of the question being asked, and the size of the budget.

In-person interviews are generally the most effective for complex questions, because it is often easier to explain the required background information to respondents in person, and people are more likely to complete a long survey when they are interviewed in person. In some cases, visual aids such as videos or colour photographs may be presented to help respondents understand the conditions of the scenario that they are being asked to value.

In-person interviews are generally the most expensive type of survey. However, mail surveys that follow procedures that aim to obtain high response rates can also be quite expensive. Mail and telephone surveys must be kept fairly short, or response rates are likely to drop dramatically. Telephone surveys may be less expensive, but it is often difficult to ask contingent valuation questions over the telephone, because of the amount of background information required.

In this hypothetical case, the researchers have decided to conduct a mail survey, because they want to survey a large sample, over a large geographical area, and are asking questions about a specific site and its benefits, which should be relatively easy to describe in writing in a relatively short survey.

Step 3

The next step is the actual survey design. This is the most important and difficult part of the process, and may take six months or more to complete. It is accomplished in several steps. The survey design process usually starts with initial interviews and/or focus groups with the types of people who will be receiving the final survey, in this case the general public. In the initial focus groups, the researchers would ask general questions, including questions about peoples' understanding of the issues related to the site, whether they are familiar with the site and its wildlife, whether and how they value this site and the habitat services it provides.

In later focus groups, the questions would get more detailed and specific, to help develop specific questions for the survey, as well as decide what kind of background information is needed and how to present it. For example, people might need information on the location and characteristics of the site, the uniqueness of species that have important habitat there, and whether there are any substitute sites that provide similar habitat. The researchers would also want to learn about peoples' knowledge of mining and its impacts, and whether mining is a controversial use of the site. If people are opposed to mining, they may answer the valuation questions with this in mind, rather than expressing their value for the services of the site. At this stage, test different approaches to the valuation question and different payment mechanisms would be tested. Questions that can identify any "protest" bids or other answers that do not reveal peoples' values for the services of interest would also be developed and tested at this stage.

After a number of focus groups have been conducted, and the researchers have reached a point where they have an idea of how to provide background information, describe the hypothetical scenario, and ask the valuation question, they will start pre-testing the survey. Because the survey will be conducted by mail, it should be pretested with as little interaction with the researchers as possible. People would be asked to assume that they've received the survey in the mail and to fill it out. Then the researchers would ask respondents about how they filled it out, and let them ask questions about anything they found confusing. Eventually, a mail pretest might be conducted. The researchers continue this process until they've developed a survey that people seem to understand and answer in a way that makes sense and reveals their values for the services of the site.

Step 4

The next step is the actual survey implementation. The first task is to select the survey sample. Ideally, the sample should be a randomly selected sample of the relevant population, using standard statistical sampling methods. In the case of a mail survey, the researchers must obtain a mailing list of randomly sampled U.S. citizens. They would then use a standard repeat-mailing and reminder method, in order to get the greatest possible response rate for the survey. Telephone surveys are carried out in a similar way, with a certain number of calls to try to reach the selected respondents. In-person surveys may be conducted with random samples of respondents, or may use "convenience" samples – asking people in public places to fill out the survey.

Step 5

The final step is to compile, analyze and report the results. The data must be entered and analyzed using statistical techniques appropriate for the type of question. In the data analysis, the researchers also attempt to identify any responses that may not express the respondent's value for the services of the site. In addition, they can deal with possible non-response bias in a number of ways. The most conservative way is to assume that those who did not respond have zero value.

How Do We Use the Results?

From the analysis, the researchers can estimate the average value for an individual or household in the sample, and extrapolate this to the relevant population in order to calculate the total benefits from the site. For example, if they find that the mean willingness to pay is $.10 per capita, the total benefits to all citizens would be $26 million.

Applying the Contingent Valuation Method

Applying the contingent valuation method is generally a complicated, lengthy, and expensive process. In order to collect useful data and provide meaningful results, the contingent valuation survey must be properly designed, pre-tested, and implemented. Contingent valuation survey questions must focus on specific environmental service(s) and a specific context that is clearly defined and understood by survey respondents. In other words, a CV survey to assess the dollar value of the results of an environmental improvement cannot be based on the environmental improvement itself, but on increases in specific environmental services that the improvement is expected to provide.

The results of contingent valuation surveys are often highly sensitive to what people believe they are being asked to value, as well as the context that is described in the survey. Thus, it is essential for CV researchers to clearly define the services and the context, and to demonstrate that respondents are actually stating their values for these services when they answer the valuation questions.

A good CV study will consider the following in its application: Before designing the survey, learn as much as possible about how people think about the good or service in question. Consider people's familiarity with the good or service, as well as the importance of such factors as quality, quantity, accessibility, the availability of substitutes, and the reversibility of the change.

Determine the extent of the affected populations or markets for the good or service in question, and choose the survey sample based on the appropriate population.

The choice scenario must provide an accurate and clear description of the change in environmental services associated with the event, programme, investment, or policy choice under consideration. If possible, convey this information using photographs, videos, or other multi-media techniques, as well as written and verbal descriptions.

Unlike ordinary survey questions, which sometimes ask respondents whether they are willing to pay x dollars to improve "air quality," the nature of the good and the changes to be valued must be specified in detail in a CV survey. It is also important to make sure that respondents do not inadvertently assume that one or more related improvements are included. For example, if people are asked to value only air visibility, it would be important to make sure that they do not include their value for health-related improvements in their stated willingness to pay amount. Similarly, if people have a tendency to think of environmental improvements in general, even when asked about water quality alone, it would be necessary to point out specifically that environmental quality, other than water quality, would remain the same.

Questions can be asked in a variety of ways, using both open-ended and closed-ended formats. In the open-ended format, respondents are asked to state their maximum willingness to pay for the environmental improvement. With the closed-ended format, also referred to as discrete choice, respondents are asked whether or not they would be willing to pay a particular amount for the environmental improvement, or whether they would vote yes or no for a specific policy at a given cost. The discrete choice format is generally accepted as the preferred method.

In addition to the hypothetical question that asks for willingness to pay, the survey must specify the mechanism by which the payment will be made, for example through increased taxes. In order for the question to be effective, the respondent must believe that if the money was paid, whoever was collecting it could effect the specified environmental change.

☆ Respondents should be reminded to consider their budget constraints.

☆ Specify whether comparable services are available from other sources, when the good is going to be provided, and whether the losses or gains are temporary or permanent.

☆ Respondents should understand the frequency of payments required, for example monthly or annually, and whether or not the payments will be required over a long period of time in order to maintain the quantity or quality change. They should also understand who would have access to the good and who else will pay for it, if it is provided.

☆ In the case of collectively held goods, respondents should understand that they are currently paying for a given level of supply. The scenario should clearly indicate whether the levels being valued are improvements over the status quo, or potential declines in the absence of sufficient payments.

☆ If the household is the unit of analysis, the reference income should be the household's, rather than the respondent's, income.

☆ Thoroughly pre-test the valuation questionnaire for potential biases. Pre-testing includes testing different ways of asking the same question, testing whether the question is sensitive to changes in the description of the good or resource being valued, and conducting post-survey interviews to determine whether respondents are stating their values as expected.

☆ Include validation questions in the survey, to verify comprehension and acceptance of the scenario, and to elicit socio-economic and attitudinal characteristics of respondents, in order to better interpret variation in responses across respondents.

☆ CVM can be conducted as in-person interviews, telephone interviews or mail surveys. The in-person interview is the most expensive survey administration format, but is generally considered to be the best approach, especially if visual materials are to be presented.

☆ Interview a large, clearly defined, representative sample of the affected population.

☆ Achieve a high response rate and a mix of respondents that represents the population.

☆ Whatever survey instruments and survey designs are used, and whatever response rate is achieved, make sure that survey results are analyzed and interpreted by professionals before making any claims about the resulting dollar values.

Advantages of the Contingent Valuation Method

☆ Contingent valuation is enormously flexible in that it can be used to estimate the economic value of virtually anything. However, it is best able to estimate values for goods and services that are easily identified and understood by users and that are consumed in discrete units (*e.g.*, user days of recreation), even if there is no observable behavior available to deduce values through other means.

☆ CV is the most widely accepted method for estimating total economic value, including all types of non-use, or "passive use," values. CV can estimate use values, as well as existence values, option values, and bequest values.

☆ Though the technique requires competent survey analysts to achieve defensible estimates, the nature of CV studies and the results of CV studies are not difficult to analyze and describe. Dollar values can be presented in terms of a mean or median value per capita or per household, or as an aggregate value for the affected population.

☆ CV has been widely used, and a great deal of research is being conducted to improve the methodology, make results more valid and reliable, and better understand its strengths and limitations.

Issues and Limitations of the Contingent Valuation Method

☆ Although the contingent valuation method has been widely used for the past two decades, there is considerable controversy over whether it adequately measures people's willingness to pay for environmental quality.

☆ People have practice making choices with market goods, so their purchasing decisions in markets are likely to reflect their true willingness to pay. CV assumes that people understand the good in question and will reveal their preferences in the contingent market just as they would in a real market. However, most people are unfamiliar with placing dollar values on environmental goods and services. Therefore, they may not have an adequate basis for stating their true value.

☆ The expressed answers to a willingness to pay question in a contingent valuation format may be biased because the respondent is actually answering a different question than the surveyor had intended. Rather than expressing value for the good, the respondent might actually be expressing their feelings about the scenario or the valuation exercise itself. For example, respondents may express a positive willingness to pay because they feel good about the act of giving for a social good (referred to as the "warm glow" effect), although they believe that the good itself is unimportant. Respondents may state a positive willingness to pay in order to signal that they place importance on improved environmental quality in general. Alternatively, some respondents may value the good, but state that they are not willing to pay for it, because they are protesting some aspect of the scenario, such as increased taxes or the means of providing the good.

☆ Respondents may make associations among environmental goods that the researcher had not intended. For example, if asked for willingness to pay for improved visibility (through reduced pollution), the respondent may actually answer based on the health risks that he or she associates with dirty air.

☆ Some researchers argue that there is a fundamental difference in the way that people make hypothetical decisions relative to the way they make actual decisions. For example, respondents may fail to take questions seriously because they will not actually be required to pay the stated amount. Responses may be unrealistically high if respondents believe they will not have to pay for the good or service and that their answer may influence the resulting supply of the good. Conversely, responses may be unrealistically low if respondents believe they will have to pay.

☆ The payment question can either be phrased as the conventional 'What are you willing to pay (WTP) to receive this environmental asset?', or in the less usual form, 'What are you willing to accept (WTA) in compensation for giving up this environmental asset?' In theory, the results should be very close. However, when the two formats have been compared, WTA very significantly exceeds WTP. Critics have claimed that this result invalidates the CVM approach, showing responses to be expressions of what individuals would like to have happen rather than true valuations.

☆ If people are first asked for their willingness to pay for one part of an environmental asset (*e.g.*, one lake in an entire system of lakes) and then asked to value the whole asset (*e.g.*, the whole lake system), the amounts stated may be similar. This is referred to as the "embedding effect."

☆ In some cases, people's expressed willingness to pay for something has been found to depend on where it is placed on a list of things being valued. This is referred to as the "ordering problem."

☆ Respondents may give different willingness to pay amounts, depending on the specific payment vehicle chosen. For example, some payment vehicles, such as taxes, may lead to protest responses from people who do not want increased taxes. Others, such as a contribution or donation, may lead people to answer in terms of how much they think their "fair share" contribution is, rather than expressing their actual value for the good.

☆ Many early studies attempted to prompt respondents by suggesting a starting bid and then increasing or decreasing this bid based upon whether the respondent agreed or refused to pay a such sum. However, it has been shown that the choice of starting bid affects respondents' final willingness to pay response.

☆ Strategic bias arises when the respondent provides a biased answer in order to influence a particular outcome. If a decision to preserve a stretch of river for fishing, for example, depends on whether or not the survey produces a sufficiently large value for fishing, the respondents who enjoy fishing may be tempted to provide an answer that ensures a high value, rather than a lower value that reflects their true valuation.

☆ Information bias may arise whenever respondents are forced to value attributes with which they have little or no experience. In such cases, the amount and type of information presented to respondents may affect their answers

☆ Non-response bias is a concern when sampling respondents, since individuals who do not respond are likely to have, on average, different values than individuals who do respond.

☆ Estimates of nonuse values are difficult to validate externally.

☆ When conducted to the exacting standards of the profession, contingent valuation methods can be very expensive and time-consuming, because of the extensive pre-testing and survey work.

☆ Many people, including jurists policy-makers, economists, and others, do not believe the results of CV.

The Contingent Choice Method

The contingent choice method is similar to contingent valuation, in that it can be used to estimate economic values for virtually any ecosystem or environmental service, and can be used to estimate non-use as well as use values. Like contingent valuation, it is a hypothetical method – it asks people to make choices based on a hypothetical scenario. However, it differs from contingent valuation because it does not directly ask people to state their values in dollars. Instead, values are inferred from the hypothetical choices or tradeoffs that people make.

The contingent choice method asks the respondent to state a preference between one group of environmental services or characteristics, at a given price or cost to the individual, and another group of environmental characteristics at a different price or cost. Because it focuses on tradeoffs among scenarios with different characteristics, contingent choice is especially suited to policy decisions where a set of possible actions might result in different impacts on natural resources or environmental services. For example, improved water quality in a lake will improve the quality of several services provided by the lake, such as drinking water supply, fishing, swimming, and biodiversity. In addition, while contingent choice can be used to estimate dollar values, the results may also be used to simply rank options, without focusing on dollar values.

This section continues with some example applications of the contingent choice method, followed by a more complete technical description of the method and its advantages and limitations.

Hypothetical Scenario

In the contingent valuation section, we used the case of a remote site on public land that provides important habitat for several species of wildlife. The management agency in charge of the area must decide whether to issue a lease for mining at the site. Suppose that there are several possible options for preserving and/or using the site. These include allowing no mining and preserving the site as a wilderness habitat area, and various levels and locations for the mining operation, each of which would have different impacts on the site. Thus, several options must be weighed in terms of costs and benefits to the public. Again, because the area is remote, few people actually visit it, or view the animals that rely on it for habitat. Therefore, non-use values are the largest component of the value for preserving the site.

Why Use the Contingent Choice Method?

The contingent choice method was selected in this case because we want to value the outcomes of several policy options, and because non-use values are important.

Alternative Approaches

Since non-use values are significant, and few people actually visit the site, other methods, such as the travel cost method, will underestimate the benefits of preserving the site. In this case, contingent valuation methods might also be used. However, because we need to value several levels of services, based on different scenarios, the survey questions might become quite complicated.

Application of the Contingent Choice Method

Because both contingent choice and contingent valuation are hypothetical survey-based methods, their application is very similar. The main differences are in the design of the valuation question(s), and the data analysis.

Step 1

The first step is to define the valuation problem. This would include determining exactly what services are being valued, and who the relevant population is. In this case, the resource to be valued is a specific site and the services it provides – primarily wildlife habitat. Because it is federally owned public land, the relevant population would be all citizens of the U.S.

Step 2

The second step is to make preliminary decisions about the survey itself, including whether it will be conducted by mail, phone or in person, how large the sample size will be, who will be surveyed, and other related questions. The answers will depend, among other things, on the importance of the valuation issue, the complexity of the question(s) being asked, and the size of the budget.

In-person interviews are generally the most effective for complex questions, because it is often easier to explain the required background information to respondents in person, and people are more likely to complete a long survey when they are interviewed in person. In some cases, visual aids such as videos or colour photographs may be presented to help respondents understand the conditions of the scenario(s) that they are being asked to value.

In-person interviews are generally the most expensive type of survey. However, mail surveys that follow procedures that aim to obtain high response rates can also be quite expensive. Mail and telephone surveys must be kept fairly short, or response rates are likely to drop dramatically. Telephone surveys are generally not appropriate for contingent choice surveys, because of the difficulty of conveying the tradeoff questions to people over the telephone.

In this hypothetical case, the researchers have decided to conduct a mail survey, because they want to survey a large sample, over a large geographical area, and are asking questions about a specific site and its benefits, which should be relatively easy to describe in writing in a relatively short survey.

Step 3

The next step is the actual survey design. This is the most important and difficult part of the process, and may take six months or more to complete. It is accomplished in several steps. The survey design process usually starts with initial interviews and/or focus groups with the types of people who will be receiving the final survey, in this case the general public. In the initial focus groups, the researchers would ask general questions, including questions about peoples' understanding of the issues related to the site, whether they are familiar with the site and its wildlife, whether and how they value this site and the habitat services it provides.

In later focus groups, the questions would get more detailed and specific, to help develop specific questions for the survey, as well as decide what kind of background information is needed and how to present it. For example, people might need information on the location and characteristics of the site, the uniqueness of species that have important habitat there, and whether there are any substitute sites that provide similar habitat.

At this stage, the researchers would test different approaches to the choice question. Usually, a contingent choice survey will ask each respondent a series of choice questions, each presenting different combinations and levels of the relevant services, as well as the cost to the respondent of the action or policy. In this example, each choice might be described in terms of the site's ability to support each of the important wildlife species. Thus, people will be making tradeoffs among the different species that might be affected in different ways by each possible choice of scenario.

After a number of focus groups have been conducted, and the researchers have reached a point where they have an idea of how to provide background information, describe the hypothetical scenario, and ask the choice question, they will start pre-testing the survey. Because the survey will be conducted by mail, it should be pretested with as little interaction with the researchers as possible. People would be asked to assume that they've received the survey in the mail and to fill it out. Then the researchers would ask respondents about how they filled it out, and let them ask questions about anything they found confusing. Eventually, a mail pretest might be conducted. The researchers continue this process until they've developed a survey that people seem to understand and answer in a way that makes sense and reveals their values for the services of the site.

Step 4

The next step is the actual survey implementation. The first task is to select the survey sample. Ideally, the sample should be a randomly selected sample of the relevant population, using standard statistical sampling methods. In the case of a mail survey, the researchers must obtain a mailing list of randomly sampled U.S. citizens. They would then use a standard repeat-mailing and reminder method, in order to get the greatest possible response rate for the survey. Telephone surveys are carried out in a similar way, with a certain number of calls to try to reach the selected respondents. In-person surveys may be conducted with random samples of respondents, or may use "convenience" samples – asking people in public places to fill out the survey.

Step 5

The final step is to compile, analyze and report the results. The statistical analysis for contingent choice is often more complicated than that for contingent valuation, requiring the use of discrete choice analysis methods to infer willingness to pay from the tradeoffs made by respondents.

From the analysis, the researchers can estimate the average value for each of the services of the site, for an individual or household in our sample. This can be extrapolated to the relevant population in order to calculate the total benefits from the site under different policy scenarios. The average value for a specific action and its outcomes can also be estimated, or the different policy options can simply be ranked in terms of peoples' preferences.

How Do We Use the Results

The results of the survey might show that the economic benefits of preserving the site by not allowing mining are greater than the benefits received from allowing mining. If this were the case, the mining lease might not be issued, unless other factors override these results. Alternatively, the results might indicate that some mining scenarios are acceptable, in terms of economic costs and benefits. The results could then be used to rank different options, and to help select the most preferred option.

Applying the Contingent Choice Method

There are a variety of formats for applying contingent choice methods, including:

☆ **Contingent Ranking**–Contingent ranking surveys ask individuals to compare and rank alternate program outcomes with various characteristics, including costs. For instance, people might be asked to compare and rank several mutually exclusive environmental improvement programs under consideration for a watershed, each of which has different outcomes and different costs. Respondents are asked to rank the alternatives in order of preference.

☆ **Discrete Choice**–In the discrete choice approach, respondents are simultaneously shown two or more different alternatives and their characteristics, and asked to identify the most preferred alternative in the choice.

☆ **Paired Rating**–This is a variation on the discrete choice format, where respondents are asked to compare two alternate situations and are asked to rate them in terms of strength of preference. For instance, people might be asked to compare two environmental improvement programs and their outcomes, and state which is preferred, and whether it is strongly, moderately, or slightly preferred to the other programme.

Whatever format is selected, the choices that respondents make are statistically analyzed using discrete choice statistical techniques, to determine the relative values for the different characteristics or attributes. If one of the characteristics is a monetary price, then it is possible to compute the respondent's willingness to pay for the other characteristics.

As with contingent valuation, in order to collect useful data and provide meaningful results, the contingent choice survey must be properly designed, pre-tested, and implemented. However, because responses are focused on tradeoffs, rather than direct expressions of dollar values, contingent choice may minimize some of the problems associated with contingent valuation. Often, relative values are easier and more natural for people to express than absolute values.

As with contingent valuation, a good contingent choice study will consider the following in its application:

☆ Before designing the survey, learn as much as possible about how people think about the good or service in question. Consider people's familiarity with the good or service, as well as the importance of such factors as quality, quantity, accessibility, the availability of substitutes, and the reversibility of the change.

☆ Determine the extent of the affected populations or markets for the good or service in question, and choose the survey sample based on the appropriate population.

☆ The choice scenario must provide an accurate and clear description of the change in environmental services associated with the event, programme, investment, or policy choice under consideration. If possible, convey this information using photographs, videos, or other multi-media techniques, as well as written and verbal descriptions.

☆ The nature of the good and the changes to be valued must be specified in detail, and it is important to make sure that respondents do not inadvertently assume that one or more related improvements are included.

☆ The respondent must believe that if the money was paid, whoever was collecting it could effect the specified environmental change.

☆ Respondents should be reminded to consider their budget constraints.

☆ Specify whether comparable services are available from other sources, when the good is going to be provided, and whether the losses or gains are temporary or permanent.

☆ Respondents should understand the frequency of payments required, for example monthly or annually, and whether or not the payments will be required over a long period of time in order to maintain the quantity or quality change. They should also understand who would have access to the good and who else will pay for it, if it is provided.

☆ In the case of collectively held goods, respondents should understand that they are currently paying for a given level of supply. The scenario should clearly indicate whether the levels being valued are improvements over the status quo, or potential declines in the absence of sufficient payments.

☆ If the household is the unit of analysis, the reference income should be the household's, rather than the respondent's, income.

☆ Thoroughly pre-test the questionnaire for potential biases. Pre-testing includes testing different ways of asking the same question, testing whether the question is sensitive to changes in the description of the good or resource being valued, and conducting post-survey interviews to determine whether respondents are stating their values as expected.

☆ Include validation questions in the survey, to verify comprehension and acceptance of the scenario, and to elicit socio-economic and attitudinal characteristics of respondents, in order to better interpret variation in responses across respondents.

☆ Surveys can be conducted as in-person interviews, telephone interviews or mail surveys. The in-person interview is the most expensive survey administration format, but is generally considered to be the best approach, especially if visual materials are to be presented. Telephone surveys are generally not effective for presenting contingent choice questions.

☆ Interview a large, clearly defined, representative sample of the affected population.

☆ Achieve a high response rate and a mix of respondents that represents the population.

☆ Whatever survey instruments and survey designs are used, and whatever response rate is achieved, make sure that survey results are analyzed and interpreted by professionals before making any claims about the resulting dollar values.

Advantages of the Contingent Choice Method

☆ The contingent choice method can be used to value the outcomes of an action as a whole, as well as the various attributes or effects of the action.

☆ The method allows respondents to think in terms of trade offs, which may be easier than directly expressing dollar values. The tradeoff process may encourage respondent introspection and make it easier to check for consistency of responses. In addition, respondents may be able to give more meaningful answers to questions about their behavior (*i.e.*, they prefer one alternative over another), than to questions that ask them directly about the dollar value of a good or service or the value of changes in environmental quality. Thus, an advantage of this method over the contingent valuation method is that it does not ask the respondent to make a tradeoff *directly* between environmental quality and money.

☆ Respondents are generally more comfortable providing qualitative rankings or ratings of attribute bundles that include prices, rather than dollar valuation of the same bundles without prices, by de-emphasizing price as simply another attribute.

☆ Survey methods may be better at estimating relative values than absolute values. Thus, even if the absolute dollar values estimated are not precise, the relative values or priorities elicited by a contingent choice survey are likely to be valid and useful for policy decisions.

☆ The method minimizes many of the biases that can arise in open-ended contingent valuation studies where respondents are presented with the unfamiliar and often unrealistic task of putting prices on non-market amenities.

☆ The method has the potential to reduce problems such as expressions of symbolic values, protest bids, and some of the other sources of potential bias associated with contingent valuation.

Issues and Limitations of the Contingent Choice Method

☆ Respondents may find some tradeoffs difficult to evaluate, because they are unfamiliar.

☆ The respondents' behaviour underlying the results of a contingent choice study is not well understood. Respondents may resort to simplified decision rules if the choices are too complicated, which can bias the results of the statistical analysis.

☆ If the number of attributes or levels of attributes is increased, the sample size and/or number of comparisons each respondent makes must be increased.

☆ When presented with a large number of tradeoff questions, respondents may lose interest or become frustrated.

☆ Contingent choice may extract preferences in the form of attitudes instead of behavior intentions.

☆ By only providing a limited number of options, it may force respondents to make choices that they would not voluntarily make.

☆ Contingent ranking requires more sophisticated statistical techniques to estimate willingness to pay.

☆ Translating the answers into dollar values, may lead to greater uncertainty in the actual value that is placed on the good or service of interest.

☆ Although contingent choice has been widely used in the field of market research, its validity and reliability for valuing non-market commodities is largely untested.

The Travel Cost Method

The travel cost method is used to estimate economic use values associated with ecosystems or sites that are used for recreation.

The method can be used to estimate the economic benefits or costs resulting from:

☆ changes in access costs for a recreational site

☆ elimination of an existing recreational site

☆ addition of a new recreational site

☆ changes in environmental quality at a recreational site

The basic premise of the travel cost method is that the time and travel cost expenses that people incur to visit a site represent the "price" of access to the site. Thus, peoples' willingness to pay to visit the site can be estimated based on the number of trips that they make at different travel costs. This is analogous to estimating peoples' willingness to pay for a marketed good based on the quantity demanded at different prices.

This section continues with some example applications of the travel cost method, followed by a more complete technical description of the method and its advantages and limitations.

Hypothetical Situation

A site used mainly for recreational fishing is threatened by development in the surrounding area. Pollution and other impacts from this development could destroy the fish habitat at the site, resulting in a serious decline in, or total loss of, the site's ability to provide recreational fishing services. Resource agency staff want to determine the value of programs or actions to protect fish habitat at the site.

Why Use the Travel Cost Method?

The travel cost method was selected in this case for two main reasons:

1. The site is primarily valuable to people as a recreational site. There are no endangered species or other highly unique qualities that would make non-use values for the site significant.

2. The expenditures for projects to protect the site are relatively low. Thus, using a relatively inexpensive method like travel cost makes the most sense.

Alternative Approaches

Contingent valuation or contingent choice methods could also be used in this case. While they might produce more precise estimates of values for specific characteristics of the site, and also could capture non-use values, they would be considerably more complicated and expensive to apply.

Options for Applying the Travel Cost Method

There are several ways to approach the problem, using variations of the travel cost method.

These include:

1. A simple zonal travel cost approach, using mostly secondary data, with some simple data collected from visitors.

2. An individual travel cost approach, using a more detailed survey of visitors.

3. A random utility approach using survey and other data, and more complicated statistical techniques.

Application of the Zonal Travel Cost Approach

The zonal travel cost method is the simplest and least expensive approach. It will estimate a value for recreational services of the site as a whole. It cannot easily be used to value a change in quality of recreation for a site, and may not consider some of the factors that may be important determinants of value.

The zonal travel cost method is applied by collecting information on the number of visits to the site from different distances. Because the travel and time costs will increase with distance, this information allows the researcher to calculate the number of visits "purchased" at different "prices." This information is used to construct the demand function for the site, and estimate the consumer surplus, or economic benefits, for the recreational services of the site.

Step 1

The first step is to define a set of zones surrounding the site. These may be defined by concentric circles around the site, or by geographic divisions that make sense, such as metropolitan areas or counties surrounding the site at different distances.

Step 2

The second step is to collect information on the number of visitors from each zone, and the number of visits made in the last year. For this hypothetical example, assume that staff at the site keep records of the number of visitors and their zipcode, which can be used to calculate total visits per zone over the last year.

Step 3

The third step is to calculate the visitation rates per 1000 population in each zone. This is simply the total visits per year from the zone, divided by the zone's population in thousands. An example is shown in the Table 1.

Table 1

Zone	Total Visits/Year	Zone Population	Visits/1000
0	400	1000	400
1	400	2000	200
2	400	4000	100
3	400	8000	50
Beyond 3	0		
Total Visits	1600		

Step 4

The fourth step is to calculate the average round-trip travel distance and travel time to the site for each zone. Assume that people in Zone 0 have zero travel distance and time. Each other zone will have an increasing travel time and distance. Next, using average cost per mile and per hour of travel time, the researcher can calculate the travel cost per trip. A standard cost per mile for operating an automobile is readily available from AAA or other sources. Assume that this cost per mile is $.30. The cost of time is more complicated. The simplest approach is to use the average hourly wage. Assume that it is $9/hour, or $.15/minute, for all zones, although in practice it is likely to differ by zone. The calculations are shown in the Table 2.

Table 2

Zone	Round Trip Travel Distance	Travel Time Round Trip	Distance Times Cost/Mile ($.30)	Travel Time Times Cost/ Minute ($.15)	Total Travel Cost/Trip
0	0	0	0	0	0
1	20	30	$6	$4.50	$10.50
2	40	60	$12	$9.00	$21.00
3	80	120	$24	$18.00	$42.00

Step 5

The fifth step is to estimate, using regression analysis, the equation that relates visits per capita to travel costs and other important variables. From this, the researcher can estimate the demand function for the average visitor. In this simple model, the analysis might include demographic variables, such as age, income, gender, and education levels, using the average values for each zone. To maintain the simplest possible model, calculating the equation with only the travel cost and visits/1000, Visits/1000 = 330 – 7.755*(Travel Cost).

Step 6

The sixth step is to construct the demand function for visits to the site, using the results of the regression analysis. The first point on the demand curve is the total visitors to the site at current access costs (assuming there is no entry fee for the site), which in this example is 1600 visits per year. The other points are found by estimating the number of visitors with different hypothetical entrance fees (assuming that an entrance fee is viewed in the same way as travel costs).

For the purposes of our example, start by assuming a $10 entrance fee. Plugging this into the estimated regression equation, V = 330 – 7.755C, gives the following:

Table 3

Zone	Travel Cost plus $10	Visits/1000	Population	Total Visits
0	$10	252	1000	252
1	$20.50	171	2000	342
2	$31.00	90	4000	360
3	$52.00	0	8000	0
		Total Visits	954	

This gives the second point on the demand curve – 954 visits at an entry fee of $10. In the same way, the number of visits for increasing entry fees can be calculated, to get:

Table 4

Entry Fee	Total Visits
$20	409
$30	129
$40	20
$50	0

These points give the demand curve for trips to the site.

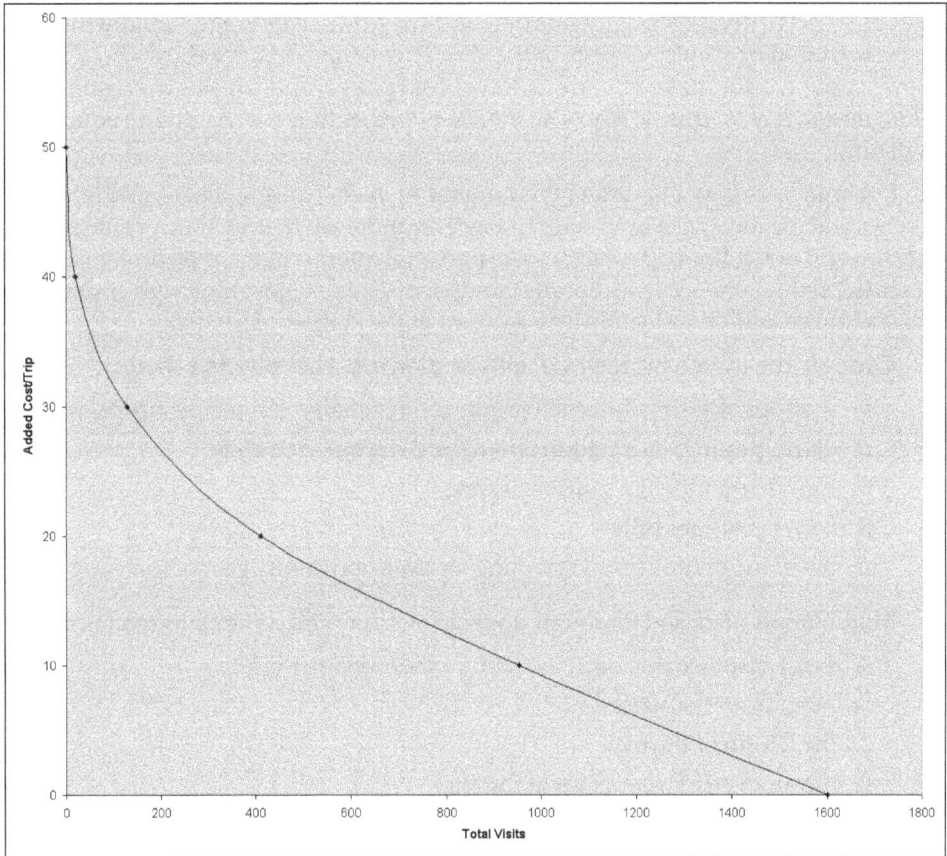

Figure 3: Zonal travel cost demand curve.

Step 7

The final step is to estimate the total economic benefit of the site to visitors by calculating the consumer surplus, or the area under the demand curve. This results in a total estimate of economic benefits from recreational uses of the site of around $23,000 per year, or around $14.38 per visit ($23,000/1,600).

How Do We Use the Results?

Remember that the agency staff's objective was to decide whether it is worthwhile to spend money on programmes and actions to protect this site. If the actions cost less than $23,000 per year, the cost will be less than the benefits provided by the site. If the costs are greater than this, the staff will have to decide whether other factors make them worthwhile.

Applying the Travel Cost Method

On average, people who live farther from a site will visit it less often, because it costs more in terms of actual travel costs and time to reach the site. The number of visits from origin zones at different distances from the site, and travel cost from each zone, are used to derive an aggregate demand curve for visits to the site, and thus for the recreational or scenic services of the site. This demand curve shows how many visits people would make at various travel cost prices, and is used to estimate the willingness to pay for people who visit the site (whether they are charged an admission fee or not).

Other factors may also affect the number of visits to a site. People with higher incomes will usually make more trips. If there are more alternative sites, or substitutes, a person will make less trips. Factors like personal interest in the type of site, or level of recreational experience will affect the number of visits. A more thorough application will take these and other factors into account in the statistical model.

To apply the travel cost method, information must be collected about:

☆ number of visits from each origin zone (usually defined by zipcode)

☆ demographic information about people from each zone

☆ round-trip mileage from each zone

☆ travel costs per mile

☆ the value of time spent traveling, or the opportunity cost of travel time

More complicated, and thorough, applications may also collect information about:

☆ exact distance that each individual travelled to the site

☆ exact travel expenses

☆ the length of the trip

☆ the amount of time spent at the site

☆ other locations visited during the same trip, and amount of time spent at each

☆ substitute sites that the person might visit instead of this site, and the travel distance to each

☆ other reasons for the trip (is the trip only to visit the site, or for several purposes)

☆ quality of the recreational experience at the site, and at other similar sites (*e.g.*, fishing success)

☆ perceptions of environmental quality at the site

☆ characteristics of the site and other, substitute, sites

This information is typically collected through surveys – on-site, telephone or mail surveys may be used. In addition, especially for simpler applications, much information may be available from state and county resource agencies, or from federal surveys, such as the National Survey of Fishing, Hunting and Wildlife Associated Recreation, published every five years by the U.S. Fish and Wildlife Service.

The most controversial aspects of the travel cost method include accounting for the opportunity cost of travel time, how to handle multi-purpose and multi-destination trips, and the fact that travel time might not be a cost to some people, but might be part of the recreational experience.

Advantages of the Travel Cost Method

☆ The travel cost method closely mimics the more conventional empirical techniques used by economists to estimate economic values based on market prices.

☆ The method is based on actual behavior – what people actually do – rather than stated willingness to pay – what people say they would do in a hypothetical situation.

☆ The method is relatively inexpensive to apply.

☆ On-site surveys provide opportunities for large sample sizes, as visitors tend to be interested in participating.

☆ The results are relatively easy to interpret and explain.

Issues and Limitations of the Travel Cost Method

☆ The travel cost method assumes that people perceive and respond to changes in travel costs the same way that they would respond to changes in admission price.

☆ The most simple models assume that individuals take a trip for a single purpose – to visit a specific recreational site. Thus, if a trip has more than one purpose, the value of the site may be overestimated. It can be difficult to apportion the travel costs among the various purposes.

☆ Defining and measuring the opportunity cost of time, or the value of time spent traveling, can be problematic. Because the time spent traveling could have been used in other ways, it has an "opportunity cost." This should be added to the travel cost, or the value of the site will be underestimated. However, there is no strong consensus on the appropriate measure – the person's wage rate, or some fraction of the wage rate – and the value chosen can have a large effect on benefit estimates. In addition, if people enjoy the travel itself, then travel time becomes a benefit, not a cost, and the value of the site will be overestimated.

☆ The availability of substitute sites will affect values. For example, if two people travel the same distance, they are assumed to have the same value. However, if one person has several substitutes available but travels to this site because it is preferred, this person's value is actually higher. Some of the more complicated models account for the availability of substitutes.

☆ Those who value certain sites may choose to live nearby. If this is the case, they will have low travel costs, but high values for the site that are not captured by the method.

☆ Interviewing visitors on-site can introduce sampling biases to the analysis.

☆ Measuring recreational quality, and relating recreational quality to environmental quality can be difficult.

☆ Standard travel cost approaches provides information about current conditions, but not about gains or losses from anticipated changes in resource conditions.

☆ In order to estimate the demand function, there needs to be enough difference between distances traveled to affect travel costs and for differences in travel costs to affect the number of trips made. Thus, it is not well suited for sites near major population centers where many visitations may be from "origin zones" that are quite close to one another.

☆ The travel cost method is limited in its scope of application because it requires user participation. It cannot be used to assign values to on-site environmental features and functions that users of the site do not find valuable. It cannot be used to value off-site values supported by the site. Most importantly, it cannot be used to measure nonuse values. Thus, sites that have unique qualities that are valued by non-users will be undervalued.

☆ As in all statistical methods, certain statistical problems can affect the results. These include choice of the functional form used to estimate the demand curve, choice of the estimating method, and choice of variables included in the model.

The Productivity Method

The productivity method, also referred to as the net factor income or derived value method, is used to estimate the economic value of ecosystem products or services that contribute to the production of commercially marketed goods. It is applied in cases where the products or services of an ecosystem are used, along with other inputs, to produce a marketed good.

For example, water quality affects the productivity of irrigated agricultural crops, or the costs of purifying municipal drinking water. Thus, the economic benefits of improved water quality can be measured by the increased revenues from greater agricultural productivity, or the decreased costs of providing clean drinking water.

This section continues with some example applications of the productivity method, followed by a more complete technical description of the method and its advantages and limitations.

Hypothetical Situation

A reservoir that provides water for a city's drinking water system is being polluted by agricultural runoff. Agency staff want to determine the economic benefits of measures to eliminate the runoff.

Why Use the Productivity Method?

The productivity method was selected because this is a straightforward case where environmental quality directly affects the cost of producing a marketed good – municipal drinking water. This example is one of the simplest cases, where cleaner water is a direct substitute for other production inputs, such as water purification chemicals and filtration.

Thus, the benefits of improved water quality can be easily related to reduced water purification costs.

Application of the Productivity Method

Step 1

The first step is to specify the production function for purified drinking water. This is the functional relationship between the inputs – water of a particular quality from the reservoir, chemicals, and filtration, and the output – pure drinking water.

Step 2

The second step is to estimate how the cost of purification changes when reservoir water quality changes, using the production function estimated in the first step. The researcher would calculate the quantities of purification chemicals and filters needed for different levels of reservoir water quality, by plugging different levels of water quality into the production function. These quantities would then be multiplied by their costs.

Step 3

The final step is to estimate the economic benefits of protecting the reservoir from runoff, in terms of reduced purification costs. For example, if all runoff is eliminated, the reservoir water will need very little treatment and the purification costs for drinking water will be minimal. This can be compared to the cost of purifying water where runoff is not controlled. The difference in purification costs is an estimate of the benefits of eliminating runoff. Similarly, the benefits for different levels of runoff reduction can be estimated. This step requires information about the projected success of actions to reduce runoff, in terms of the decrease in runoff and the resulting changes in reservoir water quality.

How Can the Results be Used

The results of the analysis can be used to compare the benefits of achieving different levels of water quality in the reservoir with the cost of programmes to reduce or eliminate the polluting runoff, and thus improve water quality.

Applying the Productivity Method

To apply the productivity method, data must be collected regarding how changes in the quantity or quality of the natural resource affect: Costs of production for the final good supply and demand for the final good and supply and demand for other factors of production.

This information is used to link the effects of changes in the quantity or quality of the resource to changes in consumer surplus and/or producer surplus, and thus to estimate the economic benefits.

The method is most easily applied in two specific cases:

1. Cases where the resource in question is a perfect substitute for other inputs. For example, increased water quality in a reservoir means that less chlorine is needed for treating the water. In this case, an increase in quantity or quality of the resource will result in decreased costs for the other inputs. Thus, in this example, the benefits of increased water quality can be directly measured by the decreased chlorination costs.

2. Cases where only producers of the final good benefit from changes in quantity or quality of the resource. Consumers are not affected. For example, improved quality of irrigation water may lead to greater agricultural productivity – more crops are produced on the same amount of land. If the market price of the crops to consumers does not change, benefits can be estimated from changes in producer surplus resulting from increased income from the other inputs. Thus, in this example, the profits per acre will increase, and this increase can be used to estimate the benefits of improved irrigation water quality.

Advantages of the Productivity Method

☆ In general, the methodology is straightforward.

☆ Data requirements are limited, and the relevant data may be readily available, so the method can be relatively inexpensive to apply.

Issues and Limitations of the Productivity Method

☆ The method is limited to valuing those resources that can be used as inputs in production of marketed goods.

☆ When valuing an ecosystem, not all services will be related to the production of marketed goods. Thus, the inferred value of that ecosystem may understate its true value to society.

☆ Information is needed on the scientific relationships between actions to improve quality or quantity of the resource and the actual outcomes of those actions. In some cases, these relationships may not be well known or understood.

☆ If the changes in the natural resource affect the market price of the final good, or the prices of any other production inputs, the method becomes much more complicated and difficult to apply.

The benefit transfer method is used to estimate economic values for ecosystem services by transferring available information from studies already completed in another location and/or context. For example, values for recreational fishing in a particular state may be estimated by applying measures of recreational fishing values from a study conducted in another state.

Thus, the basic goal of benefit transfer is to estimate benefits for one context by adapting an estimate of benefits from some other context. Benefit transfer is often used when it is too expensive and/or there is too little time available to conduct an original valuation study, yet some measure of benefits is needed. It is important to note that benefit transfers can only be as accurate as the initial study.

The Benefit Transfer Method

This section continues with some example applications of the benefit transfer method, followed by a more complete technical description of the method and its advantages and limitations.

Hypothetical Situation

A park is being upgraded to provide additional recreational opportunities. One proposal is to add a swimming beach to the lake. Agency staff want to know the benefits of the new beach, but do not want to spend a great deal of money on a valuation study.

Why Use the Benefit Transfer Method?

The benefit transfer method was selected in this case for two main reasons. First, the agency does not have a large budget for site-specific benefits studies. Second, values for recreational uses are relatively easy to transfer.

Application of the Benefit Transfer Method

Step 1

The first step is to identify existing studies or values that can be used for the transfer. In this case, the researcher would look for studies that value beach use, specifically for lake beaches if possible. For the purposes of this example, assume that the researcher has found two travel cost studies that estimated values for swimming at lake beaches.

Step 2

The second step is to decide whether the existing values are transferable. The existing values or studies would be evaluated based on several criteria, including:

1. Is the service being valued comparable to the service valued in the existing studie(s)? Some factors that determine comparability are similar types of sites (*e.g.*, lake beaches in a park), similar quality of sites (*e.g.*, water quality and facilities), and similar availability of substitutes (*e.g.*, the number of other lake beaches nearby).

2. Are characteristics of the relevant population comparable? For example, are demographics similar between the area where the existing study was conducted and the area being valued? If not, are data available to make adjustments?

In the example, the first study is for a similar lake beach. The beach is also in a park, has comparable water quality and facilities, and a similar number of substitute sites in the area. However, it is located in an urban area, while the beach being valued is in a rural area. Thus, the characteristics of visitors can be expected to be different for the two sites. The second study is in a rural area with similar types of visitors, but the lake has many more available substitutes.

Step 3

The next step is to evaluate the quality of studies to be transferred. The better the quality of the initial study, the more accurate and useful the transferred value will be. This requires the professional judgment of the researcher. In this example, the researcher has decided that both studies are acceptable in terms of quality.

Step 4

The final step is to adjust the existing values to better reflect the values for the site under consideration, using whatever information is available and relevant. The researcher may need to collect some supplemental data in order to do this well. For example, in this case, the sites valued in each of the existing studies differ from the site of interest. The researcher might adjust the values from the first study by applying demographic data to adjust for the differences in users. If the second study has a benefit function that includes the number of substitute sites, the function could be adjusted to reflect the different number of substitutes available at the site of interest.

In addition, because the beach will be new, the researcher will need to estimate how many people will use the beach. This might be accomplished by a survey of park visitors, asking whether they would use a beach on the lake, and how many times they would use it. The researcher would then multiply these visitation estimates by the value per day for beach use (adjusted for differences in population and site characteristics), to get an estimate of the economic benefits for the new beach.

Applying the Benefit Transfer Method

Application of the benefit transfer method involves several steps. First, identify existing studies or values that can be used for the transfer. There are a number of valuation databases that can be useful (see the Links section for more information).

Second, evaluate the existing values to determine whether they are appropriately transferable. Consider whether:

☆ The service being valued is comparable to the service valued in the existing studie(s). This includes determining whether the features and qualities of sites or ecosystems are similar, including the availability of substitutes.

☆ The characteristics of the relevant population are comparable. This includes determining whether the demographics, and peoples' preferences, are similar between the area where the existing study was conducted and the area being valued.

Third, evaluate the quality of studies to be transferred. The better the quality of the initial study, the more accurate and useful the transferred value will be. This step requires professional judgment of the researcher. Fourth, adjust the existing values to better reflect the values for the site under consideration, using whatever information is available and relevant. The researcher may need to collect supplemental data in order to do this well. For example, the researcher might survey key informants, talk to the investigators of the original studies, get the original data sets, or collect some primary data at the study site to use to make adjustments. Finally, estimate the total value by multiplying the transferred values by the number of affected people.

Issues and Limitations

☆ Benefit transfer may not be accurate, except for making gross estimates of recreational values, unless the sites share all of the site, location, and user specific characteristics.

☆ Good studies for the policy or issue in question may not be available.

☆ It may be difficult to track down appropriate studies, since many are not published.

☆ Reporting of existing studies may be inadequate to make the needed adjustments.

☆ Adequacy of existing studies may be difficult to assess.

☆ Extrapolation beyond the range of characteristics of the initial study is not recommended.

☆ Benefit transfers can only be as accurate as the initial value estimate.

☆ Unit value estimates can quickly become dated.

The Hedonic Pricing Method

The hedonic pricing method is used to estimate economic values for ecosystem or environmental services that directly affect market prices. It is most commonly applied to variations in housing prices that reflect the value of local environmental attributes.

It can be used to estimate economic benefits or costs associated with:

☆ Environmental quality, including air pollution, water pollution, or noise

☆ Environmental amenities, such as aesthetic views or proximity to recreational sites

The basic premise of the hedonic pricing method is that the price of a marketed good is related to its characteristics, or the services it provides. For example, the price of a car reflects the characteristics of that car – transportation, comfort, style, luxury, fuel economy, etc. Therefore, we can value the individual characteristics of a car or other good by looking at how the price people are willing to pay for it changes when the characteristics change. The hedonic pricing method is most often used to value environmental amenities that affect the price of residential properties.

This section continues with an example application of the hedonic pricing method, followed by a more complete technical description of the method and its advantages and limitations.

Hypothetical Situation

Agency staff want to measure the benefits of an open space preservation programme in a region where open land is rapidly being developed.

Why Use the Hedonic Pricing Method?

The hedonic pricing method was selected in this case because:

1. Housing prices in the area appear to be related to proximity to open space.
2. Data on real estate transactions and open space parcels are readily available, thus making this the least expensive and least complicated approach.

Alternative Approaches

If the open space of concern is used mainly for recreation, the travel cost method might be used. Alternatively, survey-based methods, like contingent valuation or contingent choice, might be used. However, these methods would generally be more difficult and expensive to apply.

Application of the Hedonic Pricing Method

Step 1

The first step is to collect data on residential property sales in the region for a specific time period (usually one year). The required data include:

- ☆ Selling prices and locations of residential properties
- ☆ Property characteristics that affect selling prices, such as lot size, number and size of rooms, and number of bathrooms
- ☆ Neighborhood characteristics that affect selling prices, such as property taxes, crime rates, and quality of schools
- ☆ Accessibility characteristics that affect prices, such as distances to work and shopping centers, and availability of public transportation
- ☆ Environmental characteristics that affect prices.

In this case, the environmental characteristic of concern is the proximity to open space. The researcher might collect data on the amount and type of open space within a given radius of each property, and might also note whether a property is directly adjacent to open space. Often, this type of data may be obtained from computer-based GIS (geographical information systems) maps. Data on housing prices and characteristics are available from municipal offices, multiple listing services, and other sources.

Step 2

Once the data are collected and compiled, the next step is to statistically estimate a function that relates property values to the property characteristics, including the distance to open space. The resulting function measures the portion of the property price that is attributable to each characteristic. Thus, the researcher can estimate the value of preserving open space by looking at how the value of the average home changes when the amount of open space nearby changes.

How Do We Use the Results?

The results can be used to evaluate agency investments in open space preservation. For example, specific parcels may be under consideration for protection. The hedonic value function can be used to determine the benefits of preserving each parcel, which can then be compared to the cost.

Applying the Hedonic Pricing Method Using Housing Prices

In general, the price of a house is related to the characteristics of the house and property itself, the characteristics of the neighborhood and community, and environmental characteristics. Thus, if non-environmental factors are controlled for, then any remaining differences in price can be attributed to differences in environmental quality. For example, if all characteristics of houses and neighbourhoods throughout an area were the same, except for the level of air pollution, then houses with better air quality would cost more. This higher price reflects the value of cleaner air to people who purchase houses in the area.

To apply the hedonic pricing method, the following information must be collected:

☆ A measure or index of the environmental amenity of interest.

☆ Cross-section and/or time-series data on property values and property and household characteristics for a well-defined market area that includes homes with different levels of environmental quality, or different distances to an environmental amenity, such as open space or the coastline.

The data are analyzed using regression analysis, which relates the price of the property to its characteristics and the environmental characteristic(s) of interest. Thus, the effects of different characteristics on price can be estimated. The regression results indicate how much property values will change for a small change in each characteristic, holding all other characteristics constant.

The analysis may be complicated by a number of factors. For example, the relationship between price and characteristics of the property may not be linear – prices may increase at an increasing or decreasing rate when characteristics change. In addition, many of the variables are likely to be correlated, so that their values change in similar ways. This can lead to understating the significance of some variables in the analysis. Thus, different functional forms and model specifications for the analysis must be considered.

Advantages of the Hedonic Pricing Method

☆ The method's main strength is that it can be used to estimate values based on actual choices.

☆ Property markets are relatively efficient in responding to information, so can be good indications of value.

☆ Property records are typically very reliable.

☆ Data on property sales and characteristics are readily available through many sources, and can be related to other secondary data sources to obtain descriptive variables for the analysis.

☆ The method is versatile, and can be adapted to consider several possible interactions between market goods and environmental quality.

Issues and Limitations

☆ The scope of environmental benefits that can be measured is limited to things that are related to housing prices.

☆ The method will only capture people's willingness to pay for perceived differences in environmental attributes, and their direct consequences. Thus, if people aren't aware of the linkages between the environmental attribute and benefits to them or their property, the value will not be reflected in home prices.

☆ The method assumes that people have the opportunity to select the combination of features they prefer, given their income. However, the housing market may be affected by outside influences, like taxes, interest rates, or other factors.

☆ The method is relatively complex to implement and interpret, requiring a high degree of statistical expertise.

☆ The results depend heavily on model specification.

☆ Large amounts of data must be gathered and manipulated.

☆ The time and expense to carry out an application depends on the availability and accessibility of data.

The Damage Cost Aoided, Replacement Cost, and Substitute Cost Methods

The damage cost avoided, replacement cost, and substitute cost methods are related methods that estimate values of ecosystem services based on either the costs of avoiding damages due to lost services, the cost of replacing ecosystem services, or the cost of providing substitute services. These methods do not provide strict measures of economic values, which are based on peoples' willingness to pay for a product or service. Instead, they assume that the costs of avoiding damages or replacing ecosystems or their services provide useful estimates of the value of these ecosystems or services. This is based on the assumption that, if people incur costs to avoid damages caused by lost ecosystem services, or to replace the services of ecosystems, then those services must be worth at least what people paid to replace them. Thus, the methods are most appropriately applied in cases where damage avoidance or replacement expenditures have actually been, or will actually be, made.

Some examples of cases where these methods might be applied include:

☆ Valuing improved water quality by measuring the cost of controlling effluent emissions.

☆ Valuing erosion protection services of a forest or wetland by measuring the cost of removing eroded sediment from downstream areas.

☆ Valuing the water purification services of a wetland by measuring the cost of filtering and chemically treating water.

☆ Valuing storm protection services of coastal wetlands by measuring the cost of building retaining walls.

☆ Valuing fish habitat and nursery services by measuring the cost of fish breeding and stocking programmes.

This section continues with some example applications of the cost-based methods, followed by a more complete technical description of the methods and their advantages and limitations.

Hypothetical Situation

An agency is considering restoring some degraded wetlands in order to improve their ability to protect the surrounding area from flooding. The agency wants to value the benefits of improved flood protection.

Why Use the Cost-Based Methods

This method was selected in this case because the agency is only interested in valuing the flood protection services of the wetlands, and they do not have a large budget available for a valuation study. A cost-based method may be the easiest and least costly method to apply in this case.

Application of the Cost-Based Methods

Step 1

The first step is to conduct an ecological assessment of the flood protection services provided by the wetlands. This assessment would determine the current level of flood protection, and the expected level of protection if the wetlands are fully restored.

Step 2

This step depends on the specific method chosen. The Damage Cost Avoided method might be applied using two different approaches. One approach is to use the information on flood protection obtained in the first step to estimate potential damages to property if flooding were to occur. In this case, the researcher would estimate, in dollars, the probable damages to property if the wetlands are not restored. A second approach would be to determine whether nearby property owners have spent money to protect their property from the possibility of flood damage, for example by purchasing additional insurance or by reinforcing their basements. These avoidance expenditures would be summed over all affected properties to provide an estimate of the benefits from increased flood protection. However, one would not expect the two approaches to produce the same estimate. One might expect that, if avoidance costs are expected to be less than the possible damages, people would pay to avoid those damages.

The replacement cost method is applied by estimating the costs of replacing the affected ecosystem services. In this case, flood protection services cannot be directly replaced, so this method would not be useful. The substitute cost method is applied by estimating the costs of providing a substitute for the affected services. For example, in this case a retaining wall or a levee might be built to protect nearby properties from flooding. The researcher would thus estimate the cost of building and maintaining such a wall or levee. The researcher must also determine whether people would be willing to accept the wall or levee in place of a restored wetland.

How Can the Results be Used

The dollar values of the property damages avoided, or of providing substitute flood protection services provide an estimate of the flood protection benefits of restoring the wetlands, and can be compared to the restoration costs to determine whether it is worthwhile to restore the flood protection services of the wetlands.

The damage cost avoided, replacement cost, and substitute cost methods are related methods that estimate values of ecosystem services based on either the costs of avoiding damages due to lost services, the cost of replacing environmental assets, or the cost of providing substitute services.

The damage cost avoided method uses either the value of property protected, or the cost of actions taken to avoid damages, as a measure of the benefits provided by an ecosystem. For example, if a wetland protects adjacent property from flooding, the flood protection benefits may be estimated by the damages avoided if the flooding does not occur or by the expenditures property owners make to protect their property from flooding.

The replacement cost method uses the cost of replacing an ecosystem or its services as an estimate of the value of the ecosystem or its services. Similarly, the substitute cost method uses the cost of providing substitutes for an ecosystem or its services as an estimate of the value of the ecosystem or its services. For example, the flood protection services of a wetland might be replaced by a retaining wall or levee.

Because these methods are based on using costs to estimate benefits, it is important to note that they do not provide a technically correct measure of economic value, which is properly measured by the maximum amount of money or other goods that a person is willing to give up to have a particular good, less the actual cost of the good. Instead, they assume that the costs of avoiding damages or replacing natural assets or their services provide useful estimates of the value of these assets or services. This is based on the assumption that, if people incur costs to avoid damages caused by lost ecosystem services, or to replace the services of ecosystems, then those services must be worth at least what people paid to replace them. This assumption may or may not be true. However, in some cases it may be reasonable to make such assumptions, and measures of damage cost avoided or replacement cost are generally much easier to estimate than people's willingness to pay for certain ecosystem services. The methods are most appropriately applied in cases where damage avoidance or replacement expenditures have actually been, or will actually be, made.

Applying the Damage Cost Avoided, Replacement Cost, and Substitute Cost Methods

These methods require the same initial step–assessing the environmental service(s) provided. This involves specifying the relevant service(s), how they are provided, to whom they are provided, and the level(s) provided. For example, in the case of flood protection, this would involve predictions of flooding occurrences and their levels, as well as the potential impacts on property.

The second step for the damage cost avoided method is to estimate the potential physical damage to property, either annually or over some discrete time period. The final step for the damage cost avoided method is to calculate either the dollar value of potential property damage, or the amount that people spend to avoid such damage.

The second step for the replacement or substitute cost method is to identify the least costly alternative means of providing the service(s). The third step is to calculate the cost of the substitute or replacement service(s). Finally, public demand for this alternative must be established. This requires gathering evidence that the public would be willing to accept the substitute or replacement service(s) in place of the ecosystem service(s).

Advantages of the Damage Cost Avoided, Replacement Cost, and Substitute Cost Methods

☆ The methods may provide a rough indicator of economic value, subject to data constraints and the degree of similarity or substitutability between related goods.

☆ It is easier to measure the costs of producing benefits than the benefits themselves, when goods, services, and benefits are non-marketed. Thus, these approaches are less data and resource-intensive.

☆ Data or resource limitations may rule out valuation methods that estimate willingness to pay.

☆ The methods provide surrogate measures of value that are as consistent as possible with the economic concept of use value, for services which may be difficult to value by other means.

Issues and Limitations

☆ These approaches assume that expenditures to repair damages or to replace ecosystem services are valid measures of the benefits provided. However, costs are usually not an accurate measure of benefits.

☆ These methods do not consider social preferences for ecosystem services, or individuals' behaviour in the absence of those services. Thus, they should be used as a last resort to value ecosystem services.

☆ The methods may be inconsistent because few environmental actions and regulations are based solely on benefit-cost comparisons, particularly at the national level. Therefore, the cost of a protective action may actually exceed the benefits to society. It is also likely that the cost of actions already taken to protect an ecological resource will underestimate the benefits of a new action to improve or protect the resource.

☆ The replacement cost method requires information on the degree of substitution between the market good and the natural resource. Few environmental resources have such direct or indirect substitutes. Substitute goods are unlikely to provide the same types of benefits as the natural resource, *e.g.*, stocked salmon may not be valued as highly by anglers as wild salmon.

☆ The goods or services being replaced probably represent only a portion of the full range of services provided by the natural resource. Thus, the benefits of an action to protect or restore the ecological resource would be understated.

☆ These approaches should be used only after a project has been implemented or if society has demonstrated their willingness-to-pay for the project in some other way (*e.g.*, approved spending for the project). Otherwise there is no indication that the value of the good or service provided by the ecological resource to the affected community greater than the estimated cost of the project.

☆ Just because an ecosystem service is eliminated is no guarantee that the public would be willing to pay for the identified least cost alternative merely because it would supply the same benefit level as that service. Without evidence that the public would demand the alternative, this methodology is not an economically appropriate estimator of ecosystem service value.

Chapter 7

Natural Resource
Valuation Techniques

This section describes methods used in the valuation of natural and environmental resources and resource services. Examples are used to illustrate their potential applicability in a variety of situations, such as

land valuation upon completion of restoration activities or water valuation following the prevention of toxic discharges. The key is to estimate the demand for the beneficial uses or services that natural resources provide individuals and communities. Where markets for the resource or its services exist, assessment is relatively straightforward. An example would be a local real estate market. Observations on the number and value of transactions provide information about the people's willingness to pay for land and the quantity of land changing hands. These market data provide a means through which to deduce the market demand curve and the actual payments made during a given period of time.

When a market such as this exists, it is relatively easy to apply market based techniques to measure value. These techniques include the market price approach, the appraisal method, and the replacement cost method. Otherwise, when market data is not available, valuation requires the use of non-market techniques to derive information on individual willingness to pay. The most widely recognized non-market techniques include the *travel cost method*, the *hedonic price method*, and the *contingent valuation method*. Also, cross-cutting methods have been used as a way to combine market-based and no market methods of valuation, such as the *benefit-transfer method* or the *unit-day value method*. Finally, other recent approaches have focused on the valuation of ecological functions. Table 1 provides an overview of these techniques and the types of benefits that are examined below. Specific consideration is given to each of these valuation techniques with the aid of box illustrations of selected case studies.

Table 1: Valuation Techniques, Benefit Types, and Selected Case Studies.

Economic Valuation Techniques	Types of Benefits	Selected Case Studies
Market price approach	Recreational/existence value	Loomis and Anderson (1992)
Appraisal methods	"Fair market values" of land	Scott *et al.* (1997)
Resource replacement cost	Groundwater resource values	Shechter (1985)
Travel cost method	Recreational/existence value	Scott *et al.* (1997)
Random utility models	Recreational/existence value	Morey *et al.* (1991)
Hedonic price method	Groundwater/land value	Kopp and Smith (1992)
	Human health/value of life	Viscusi (1990)
Factor income approach	Freshwater supply	Non considered
Contingent valuation method	Use/non-use values	Kopp and Smith (1992)
Benefit-transfer method	Air quality/visibility	Ulibarri and Ghosh (1995)
Unit-day value method	Recreational value	Loomis and Anderson (1992)
Ecological valuation approach	Gross primary energy value and intrinsic value	Constanza *et al.* (1989) and Scott *et al.* (1997)

The natural resource valuation techniques identified in Table 1 provide a relatively broad picture of the economic thinking that goes into the monetary valuation of natural resources. However, a further point should be made. Despite considerable progress over the last twenty some years, the monetary valuation of natural resources (or environmental commodities) remains in a state of flux. The natural resource valuation techniques considered in this handbook are no exception. Thus, although monetary estimates of natural resource values are given, they should be regarded as approximations – at best, an order-of-magnitude indication of the actual numbers. Unfortunately, there is hardly any research to refer to which has attempted to validate or compare the monetary value estimates identified in this handbook. As such, it is difficult to measure the range of uncertainties that underlie these estimates. For this reason, the authors have not adjusted any of the monetary values for inflation; for to do so would, in the words of Allen Kneese, "confer on them an unfounded degree of accuracy."

Market-Based Techniques

The pioneers of natural and environmental resource valuation relied on the "law of demand" as a way to measure the market values for natural resources and environmental amenities. While the same is true today, the degree of sophistication in the measurement of these values has increased considerably. Three market-based techniques that have recorded a significant history of natural and environmental resource valuations are described here: the market price approach, the appraisal method, and resource replacement costing.

Market Price Approach

Demand for natural resources is measured on the assumption that any factors that might influence demand, such as personal income, the prices of related goods and services, and individual tastes and preferences, remain unchanged during the study period. Under these assumptions, the estimated demand curve is a systematic measure of how people value the resource. To illustrate, Figure 1 shows that 20,000 acres of land were sold at a market price of $1500 per acre. In the course of these land transactions, $30.0 million exchanged hands in the land market, *i.e.*, 20,000 x $1500. Had land become increasingly scarce, this scarcity would ultimately be reflected in higher land prices. Either fewer people would purchase the land, or the same people would purchase less land.

Now, consider the total area beneath the demand curve up to 20,000 acres, as defined by A+B. This area measures the value of the resource in terms of the maximum willingness to pay for the 20,000 acres of land. The total willingness to pay for 20,000 acres is calculated by adding up what was actually spent in buying the land, A = $30 million, plus the additional triangular area B, which defines *consumer surplus*. Consumer surplus is the difference between people's maximum willingness to pay for 20,000 acres of land (A+B) and what they actually paid (A). In essence, the area gives a dollar measure of satisfaction that people received from the land, less what they actually pay for it.

As a dollar measure of individual welfare, consumer surplus and expected consumer surplus are satisfactory for most studies, and many analysts have found them to be good empirical approximations of more theoretically desirable measures. Evaluating consumer surplus requires data of market transactions for varying prices and quantities, as well as information on personal income and the prices of related goods and services. People's expenditures on resources would be an inappropriate measure of willingness to pay because it omits the consumer's surplus from the overall valuation.

Producer surplus and economic rent are two other measures of the benefits (or damages) associated with natural resources and resource services. Producer surplus measures monetary gains from the production of natural resources, which is the difference between revenues (C+D) and the economic costs of producing these resources (D). Similarly, economic rent measures monetary gains from using natural resources as factors of production, which is the difference between the actual payments made in using resources and the lowest payment that their owners would have been willing to accept in supplying these resources or resource services. Thus, producer surplus refers to the sellers' gains from trade in the product market, while economic rent measures the sellers' gains from trade in the input market. Accordingly, the use of producer surplus or economic rent in resource valuation problems depends on whether the natural resource is considered as a final product or as an input in the production of a final product.

Referring again to Figure 1, producer surplus is shown by the area C, which is bordered by the resource supply curve and the market price of the resource, P = $1500. This measure reflects changes in the availability of the natural resource. For

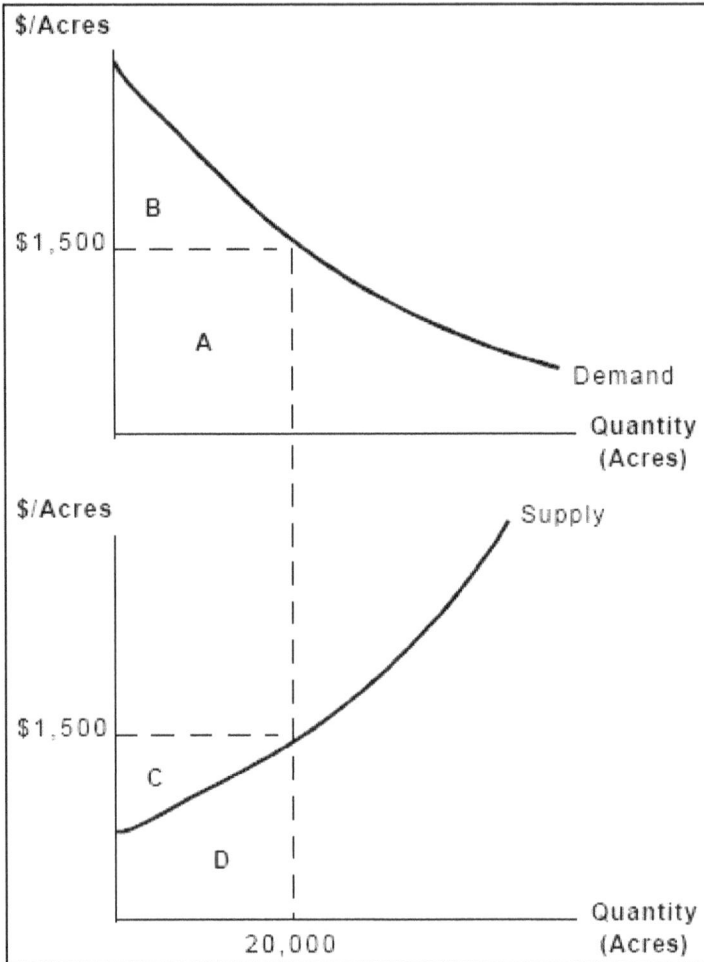

Figure 1 Demand, Supply, and Market Valuation

example, if the natural resource were damaged, its supply curve would shift leftward and producer surplus would diminish. A similar description could be given to natural resource damages that result in a reduction in economic rent. Here, the damages would be incurred by the owners of the resources. As in the case of measuring the consumer surplus, both producer surplus and economic rent require historical information on the market prices and quantities of natural resources. In addition, the measures of producer surplus and economic rent require information relating to the economic costs of producing and/or supplying the resource to the market.

Appraisal Method

Appraisal methods are particularly well suited to cases involving natural resources that have been damaged. In the case of land, for example, the appraiser identifies the fair market value for comparable properties in both the uninjured and injured conditions. The fair market value of the resource (land) is roughly defined as the amount a knowledgeable buyer would pay a knowledgeable seller for the resources. This value should reflect, as closely as possible, the price at which the resource would actually sell in the market place at the time of the injury.

The application of appraisal methods would seem to hold particular promise in DOE natural and environmental resource planning and guidance. Appraisal methods for resource valuation work have been found to be reliable under the Department of Interior's and the National Oceanic and Atmospheric Administration's natural resource damage assessment regulations. However, the point to keep in mind is that the method is, in fact, quite dependent on the appraiser's judgment. It may be very difficult to identify comparable sales, particularly for properties that are "comparably" injured. In addition, the types of natural resources to which this method can be applied are limited since many natural and environmental resources are not traded in markets. Nevertheless, appraisal methods are applicable to soil and water treatment at federal facilities. Therefore, it is instructive to consider a notable protocol in applying appraisal methods.

Scott *et al.* (1997) estimated the "fair market value" associated with shrub-steppe conversions based on sample data from Benton-Franklin Counties of eastern Washington State. The data were obtained from the Benton County Assessor's Office and represent sales transactions in Benton County involving 7700 acres during the 1993- 1994 calendar year. The sample was selected to ensure the identification of recent patterns in the regional development of shrub-steppe land. Consequently, the sample contained 17 transactions of property for residential and/or commercial development (urban use) and 31 transactions involving property destined for agricultural development (agricultural use). The authors categorized the sales of predisposed agricultural land according to whether it was irrigated, or whether it would be used as dry pasture land or dry farm land. The sampling of real estate transactions found that shrub steppe for urban development had the highest average value, $9208 per acre. Dry pasture land had the lowest average value, $67 per acre. Meanwhile, irrigated farm land sold for $1484 per acre.

Resource Replacement Cost Method

The costs of replacing natural and environmental resources are sometimes a useful way of approximating resource values under specific conditions. The resource replacement cost method determines damages for natural resources based on the cost to restore, rehabilitate, or replace the resource or resource services without injury to the level of the resource stock or service flow. In instances where the underlying resource is not unique and substitutes are readily available, the application of the replacement cost method is relatively straightforward. The investigator proceeds by gathering a sample of values for the substitutes from primary or secondary source information. Based on this sample of cost information, the analyst then prepares an estimate of the most likely range of expected replacement costs for the underlying resource or service. This process may be far more difficult to implement in instances where resources possess unique characteristics. In these cases, little information exists to assemble a sample upon which to estimate the expected value of the underlying resource.

Shechter (1985) applied the replacement cost method at the Price Landfill in New Jersey to obtain cost estimates of alternatives to deal with groundwater contamination. Estimates were based on information obtained from the U.S. Environmental Protection Agency (1978) and Environmental Science and Technology (1980). Excluding excavation and reburials, the estimated costs ranged from $5 million to $8 million (in 1980 dollars) and included containment and management of the plume, along with the performance of water treatment until the aquifer had been purged of noxious substances. If excavation and reburial were undertaken as part of the restoration process, the researchers suggest that the period of plume management and groundwater flow control could be shortened, but that total cost would rise by about $15 million to $18 million. Other site restoration activities included in their estimation focused on securing alternative sources of water to meet Atlantic City's water demand for the foreseeable future. These included cost estimates for the development of a well field to replace four threatened wells, varying between $6.5 million and $9.3 million. The researchers omitted other administrative costs from consideration in applying the method, such as the costs of undertaking various federal, state, and local studies on the landfill problem, and the attendant litigation costs that might be involved. It was believed that these administrative costs had the potential to raise the total cost by another $1.5 million.

While the replacement cost approach has been used in court settlements for damaged resources, there are problems concerning the interpretation of its meaning. For example, resource replacement cost can be viewed as merely a convenient measure for compensation without implying actual restoration of the natural resource to its previous state. Alternatively, it can be viewed as including the costs of actually restoring the natural resource habitat to its previous state and then replacing damaged organisms. Another disadvantage of the replacement cost method is that it is argued to be an arbitrary valuation of natural resources that may bear little relationship to true social value. The resource replacement cost method requires data on the costs to restore, rehabilitate, or replace injured or lost resources and resource services.

Non Market Valuation: Indirect Techniques

Using market-based techniques to measure the monetary value of natural resources is feasible provided there is sufficient market data. In many cases, however, market information relating to prices and quantities is not available to estimate the value of the resource or resource service. In these cases, researchers must employ what are referred to as non market valuation methods. These methods include indirect techniques that rely on observable behaviour in order to deduce how much something is worth to individuals. Value estimates obtained using indirect non market valuation techniques are conceptually identical to the otherwise unobservable market value. The indirect non market valuation techniques considered in this section include the travel cost method, the random utility method, the hedonic pricing method, and the factor income method.

Travel Cost Method

The travel cost method is popular for describing the demand for the natural resource service(s) and environmental attributes of specific recreational sites. Designated wilderness areas, ecological parks, fishing and hunting sites, and scenic sites are examples. People visit such sites from diverse distances or points of origin. This observed "travel behaviour" is then used to evaluate the willingness to pay to visit the site; essentially, the different travel costs from these diverse points of origin serve as proxies for willingness to pay to visit the site. Intuitively, one would expect that the environmental attributes of sites influence the use of these sites. As such, changes in visitation rates may reflect changes in the quality of natural resources particular to the site, thereby providing an estimate of the value of changes in natural resource and environmental quality.

By gathering information on the number of visits to a particular site, the analyst can estimate a demand function for the site that relates the number of site visitations to the amount of travel costs incurred per visit, taking into consideration a set of independent household variables. If first-hand information on individual visitation rates is not available to the analyst, users of the site can often be grouped into travel zones around a site. Variations in visitation rates across zones can then be used to estimate the site demand function. In this way, travel cost models provide benefit measures for changes in environmental quality found at sites, based on the observed behavior of recreational site users. Among the key advantages of applying the method at DOE sites is its adaptability to many environmental quality issues where changes in quality affect the desirability of potential recreation sites.

In addition, the travel cost method can be easily implemented using phone, onsite or mail surveys, or site registration data. In some cases, survey data may be available from local, state, and federal resource management agencies to obtain travel cost estimates of site values. The technique is generally not perceived as being particularly controversial, partly because of its long history in forestry economics, but mostly because it mimics common empirical techniques used elsewhere in economics. Analysts have tended to look favorably on the travel cost approach to natural resource valuation because it is based on actual behaviour rather than verbal responses to hypothetical scenarios. Individuals are actually observed spending money and time, and their economic values are deduced from their behaviour. In appropriate circumstances, travel cost models can often be applied without enormous expense.

The greatest disadvantage of travel cost and other indirect techniques is that they cannot be used unless there is some easily observable behavior that can be used to reveal values. In addition, travel cost models can be technically and statistically complicated. Data must be employed to statistically estimate increasingly sophisticated econometric models that take into account sample selection problems and nonlinear consumer surplus estimates. In addition, the resulting estimates sometimes have been found to be rather sensitive to arbitrary choices of the functional form of the estimating equation, the treatment of the value of an individual's time, the existence of multiple stops during the travel period, and the recognition of substitute sites.

Finally, the travel cost approach requires that the analyst be in a position to correlate environmental changes with the behaviour of visitors.

Random Utility Models

Random utility models are conceptually linked with the travel cost models in that they seek the same sorts of values and use the same sort of logic. However, random utility models provide a different structure in which to model recreational demand, one which focuses attention on choices among substitute sites for any given recreational trip instead of the number of trips taken to a given site. These models are especially suitable when substitution among quality-differentiated sites is a predominant characteristic of the problem. That is, this type of model is particularly appropriate when there are many substitutes available to the individual and when the change being valued is a change in the quality characteristics of one or more site alternatives.

Travel Cost Model

Scott *et al.* (1997) used the travel cost method to estimate willingness to pay for upland bird hunting in Benton-Franklin Counties in eastern Washington state. Valuation data were obtained from the Washington State Department of Fish and Wildlife (Upland Game Division) and the "1991 Washington Survey of Fishing, Hunting, and Wildlife Associated Recreation." The authors apportioned travel by upland bird hunters to Benton-Franklin Counties into five zones based on state averages: those that travelled less than 25 miles to their hunting site, between 25 and 50 miles, between 50 and 100 miles, between 100 and 250 miles, and over 250 miles.

The authors estimated that the average cost per small game hunter in 1991 was $193, and assumed that this cost varied in proportion to distance travelled to Benton-Franklin Counties. Given the latter assumption, they estimated an average cost per zone by multiplying the average cost of $193 by the ratio of the median distance in each zone to the average distance travelled. Using this estimation of travel costs, willingness to pay for hunting shrub-steppe dependent game birds was estimated for the individual hunting zones and then aggregated across zones to obtain a willingness-to-pay estimate of $3.2 million in annual recreational benefits.

Random Utility Modeling

Morey *et al.* (1991) considered the demand for, and benefits from, marine recreational fishing along the Oregon coast. The study estimated consumer surplus of different individuals relative to changes in species availability (particularly salmon) due to changes in ecological conditions in the Columbia River. The data for this analysis were obtained from a 1981 National Marine Fisheries Service intercept survey along the Pacific coast. Anglers were interviewed at numerous fishing sites along the Oregon coast. Information was collected about their trip and their catch, but not their distribution of trips across sites. Other angler-specific information collected included county of residence, expense of the trip, and total number of trips during the last 12 months. The average per-trip costs (travel cost plus the value of time) in the sample of anglers varied from $4.83 to $329.24, depending on the county of origin and their final destination. The authors reported considerable variation in catch rates across sites, modes of fishing, and fish species. Many of the catch rates were assumed to equal zero because not all species are available at the various fishing sites. The largest catch rate was reported to be 6.85 for rockfish from charter boats in Coos Bay.

Morey *et al.* used a discrete-choice random utility model to estimate the number of times an individual will participate in a given type of site-specific activity and which site will be selected on each trip, given different supply conditions for the natural resource. The individual consumer's surplus was measured by the "*ex-ante* seasonal compensating variation," which reflected changes in such characteristics as personal household income. For example, the consumer's surplus from visits to Clatsop County was associated with the elimination of either on-shore, off-shore, or all fishing opportunities in the county. Each individual's measure of consumer's surplus for the fishing season was then obtained by multiplying their seasonal consumer's surplus by the estimated number of seasons.

The study found that an angler from Clatsop County will pay $111.62 before the season starts to be able to fish from an on-shore mode in Clatsop throughout the season. An angler from Tillamook will pay $67.52 for the same option, but an angler from Curry County will pay only $5.88. These differences in value illustrate that an angler will pay for the opportunity of fishing at a site/mode that he/she might actually never visit, but the amount is small unless there is a significant probability that the angler will visit that site/mode. Multiplying each individual compensating variation by the number of anglers in a county and summing across counties gives an estimated aggregate yearly compensating variation of $4.2 million for the elimination of all the modes in Clatsop County. This is an estimate of how much all of the anglers in Oregon would have paid for the option of going fishing in Clatsop County in 1981.

Unlike travel cost models, however, random utility models cannot explain the total number of trips an individual takes to a given site in a season. Nonetheless, random utility models would seem to provide a useful technique for comparing benefits of site restoration or decontamination activities across waste sites at federal facilities. To the best of our knowledge, no such applications have been undertaken.

Hedonic Price Method–Amenity Value

Hedonic pricing is a useful tool in the assessment of amenity value. Early analysis related residential property values to neighbourhood amenities. These models provided an inferential measure of people's willingness to pay for the amenity under study. The method is used mostly to estimate the willingness to pay for variations in property values due to the presence or absence of specific environmental attributes, such as air quality, noise, and panoramic vistas. By comparing the market value of two properties having different degrees of a specific attribute, analysts extract the implicit value of the attribute to property buyers and sellers. A variation on the approach is to compare the price of a single piece of property over successive sales. By correcting for other factors that might have influenced the value of the property, the analyst can isolate the implicit price of an amenity or bundle of amenities that have changed over time.

Consider the impacts of the completion of a DOE environmental restoration activity on the price of neighboring land. At one time, the proximity of the parcel of private property to an abandoned DOE waste site may have reflected the disamenities of living in a hazardous environment. Years later, upon restoration of the site, the hedonic model would suggest an implicit value for DOE investments in environmental improvements. Similar analogies can be drawn in relation to the estimation of monetary damages to natural resources from environmental disamenities, or the monetary benefits of investing in their improvement. Accordingly, hedonic pricing methods appear to be well suited to DOE planning work involving Environmental Impact Statements (EISs) and Natural Resource Damage Assessments.

However, the reader should be made aware of caveats pertaining to the values obtained from hedonic price functions. In particular, the resource values that are obtained directly from the estimated hedonic price function are subject to fairly restrictive assumptions. It may be necessary to employ additional information from multiple commodity markets relating to the resource under consideration. Overall, the resulting hedonic price will depend on the availability of market information pertaining to the resource, and the revelation of buyer and seller preferences through market behavior. Market data on property sales and characteristics are available through real estate services and municipal sources and can be readily linked with other secondary data sources. Despite these positives, a guarded interpretation of the estimated welfare changes is recommended. Estimation and interpretation of these measures can be complex and the data requirements demanding, and there is a need to control for many important socio-demographic characteristics.

Hedonic Price Method–Value of Life

Hedonic pricing methods have also been applied in the estimation of economic damages associated with occupational health and safety risks and are becoming more widely accepted in the determination of personal injury awards in liability cases. Application in this branch of the hedonic valuation literature often refers to the "value of life" or the "hedonic value of life." Clearly, there is no such thing as a unique value of life. Consequently, meaningful estimates of the hedonic value of life vary according to the specific context under consideration. For one, it must be made clear whose value is under consideration: Is it a worker who understands and accepts a health/safety risk, or is it a passer-by who is unaware of the risk but nevertheless is predisposed to some adverse health impacts? Moreover, does the hedonic value under consideration concern the prevention of adverse health consequences from a potential accident, or does it concern an after-the-fact compensation to be given to survivors of an accident? To better understand the significance of these questions, it is instructive to clarify the concepts that are involved by distinguishing between two basic hedonic damage values:

Hedonic Price Method–Amenity Value

The hedonic method was used in the Eagle Mine case (Kopp and Smith, 1992). The plaintiff/trustee, the state of Colorado, contended that operation of the Eagle Mine facility near Gilman, Colorado, resulted in release of a variety of hazardous substances into the groundwater and the Eagle River and may have affected some portions of public land adjoining the river. These effects arose primarily from the disposal of mine tailings. As a direct result of the release, the trustee contended, several services provided by the Eagle River diminished both in quality and quantity. These services included recreational activities on the river, such as fishing and boating, and recreational activities near the river, such as hiking and camping. Moreover, because of these releases in the river, the plaintiff argued, its aesthetic quality had been impaired, leading to a decline in the value of adjacent properties. Finally, some private wells used for drinking water were thought to have been contaminated.

To evaluate the natural resource damages associated with these effects, the trustee used methods based on U.S. Forest Service estimates of the values per day of alternative recreational experiences ($14 per day for water-based recreation and $9 per day for nonwater-based recreation), two contingent valuation surveys, and a hedonic property value model. The hedonic price model was based on responses to the survey of Eagle County residents who answered a question about the purchase price for their homes, which were situated within 25 miles of the Eagle Mine. A variable indicating whether the home was within six miles of the mine was used to represent the effects of the mine. The objective of this model was to obtain estimates of damage due to possible contamination of local drinking water supplies and to blowing dust from the Eagle Mine tailing piles.

Unfortunately, the hedonic technique fails to capture all aspects of this proximity to the Eagle Mine. Moreover, because the differences in property values due to proximity to the Eagle Mine represent capitalized differences in the flow of services from the injured natural resources, the results of the hedonic model represent the present value of all perceived future damages. Nonetheless, the results of the study suggest a property devaluation amounting to $24,400 for property located within six miles of the Eagle Mine. Because 500 residences were located within the six miles, the plaintiff claimed an aggregate damage estimate of $12.2 million.

Insurance value is the amount that an individual is willing to pay to ensure a preferred level of welfare, assuming (*a*) that they fully understand the risk to which they are predisposed and (*b*) that the costs of buying insurance are in perfect correspondence to the specific risk under consideration. Meanwhile, deterrence values are used by leading practitioners as the appropriate measure of compensation value that should be charged from the standpoint of the accident victim. The amounts generally exceed the insurance value, as these tend to reflect individual attitudes towards all consequences of the risk. This would include the value that the individual has attached to the risk of experiencing the injury, losing income as a result of the injury, and losing the ability to enjoy life.

Conceptually, these two hedonic value-of-life measures can be used to determine the amount of compensation required to make the accident victim(s) whole by either restoring or maintaining a benchmark preaccident level of welfare. One of the most important results identified in the literature is that workers who are predisposed to a typical occupational injury would select an amount of insurance compensation below that which would be required to completely restore their pre-accident level of welfare.

To illustrate the potential applicability of hedonic value methods, consider the accompanying example, which involves estimating the monetary benefits of meeting regulatory safety or compliance standards, such as the attainment of ALARA dose limits. In considering this example, the reader should keep in mind that there are added ambiguities in determining hedonic value-of-life estimates in the context of human health risks accompanying environmental restoration and waste management activities. More specifically, there is apt to be a varying degree of onsite and offsite uncertainty associated with waste stream characteristics, the extent of toxic discharges, or transport pathways to human receptors.

Factor Income Method

The factor income method is used as a means of valuation in applications where natural resources are used as inputs in the production of other goods and services. Accordingly, the resulting economic costs of production are an important source of information in applying the factor income approach. While the method of factor income is not as welldefined or widely referenced as the hedonic price or travel cost methodologies, it is recognized by the U.S. Department of Interior's natural resource damage assessment regulations.

There are several types of resources for which the factor income approach is potentially well-suited, including surface water and groundwater resources, forests, and commercial fisheries. Surface and groundwater resources may be inputs to irrigated agriculture, to manufacturing, or to privately owned municipal water systems. The products in these cases agricultural crops, sawlogs, manufactured goods, and municipal water may all have market prices. Similarly, commercial fishery resources fish populations or stocks are inputs to the production of a catch of saleable fish. A variation on this theme may be useful for valuing damages to water resources.

In cases involving damages to water resources that are used in production processes, for example, one might identify the incremental cost of treating water sufficiently to return it to the pre-release water quality level. For example, a manufacturer who already engages in some form of water treatment as part of its production process might experience increased treatment costs because of hazardous substance releases upstream. If all other things are unchanged (product price, the mix of inputs in the production process, output levels), then the increased cost per unit of "clean water" provides a measure of lost factor income. This approach is convenient in that the costs of treating water are separated from other production costs incurred in the manufacturing process. Similarly, the example suggests that treatment costs might be applicable to a wide variety of situations of interest to DOE field operations.

There are, however, potential problems in applying the factor income approach. First, a particular treatment option might not be the least-cost or optimal response on the part of the water-using entity. For example, it might be cheaper to change the production process, buy municipal water or otherwise obtain a different source of water, or make other changes to the equipment or materials used. In this case, changes in water treatment costs may overstate damages. Second, it is possible that other things may change, particularly price and output levels. These potential problems can complicate the analysis and require the researcher to obtain additional technical information concerning the supply and demand of the underlying resource or resource service.

Contingent Valuation

Given the potential shortcomings in applying indirect nonmarket valuation techniques, researchers have advanced the use of a more direct approach, namely contingent market valuation. Contingent market analysis has estimated a wide variety of use and nonuse values.

The most obvious way to measure nonmarket values is to ask people how much they would be willing to pay for the resource or avoid any damages that might be sustained by the resource. Alternatively, one could ask how much people would be willing to accept as compensation for damages to the resource. Measures obtained using this technique rely on people's hypothetical willingness to pay rather than actual market-information on their behaviour: hence, the term contingent valuation (CV). The contingent valuation method is a survey-based approach to the valuation of nonmarket goods and services. It uses questionnaires to elicit information about the preference-related value of the natural resource in question. The value is said to be contingent upon the existence of a hypothetical market as described in the survey put to respondents. In principle, contingent valuation could be used to estimate the economic value of almost anything. By default, it is the only method that holds the promise of measuring nonuse values since all other methods depend on observing actual behavior associated with the natural resource.

Contingent valuation surveys may be conducted as face-to-face interviews, telephone interviews, or mail surveys based on a randomly selected sample or stratified sample of individuals. Face-to-face interviews are the most expensive survey administration format, but they are generally considered the best, especially if visual material needs to be presented. The central goal of the survey is to generate data on respondents' willingness to pay for (or willingness to accept) some programme or plan that will impact their well-being.

Each respondent is given information about a particular problem. Each is then presented with a hypothetical occurrence (*e.g.*, specie endangerment) or a policy action that ensures against the disaster (*e.g.*, specie protection). Each respondent is asked how much he/she would be willing to pay either to avoid the negative occurrence or bring about the positive occurrence. The means of payment (*i.e.*, the payment vehicle) can take on any number of different forms, including a direct tax, an income tax, or an access fee. The actual format may take the form of a direct question ("how much?"), a bidding procedure (a ranking of alternatives), or referenda votes. Using a referendum to elicit values is preferred because it is the one that people are most familiar with. Resulting data are then analyzed statistically and extrapolated to the population that the sample represents. These responses are gathered along with socio-demographic information and test statistics required to determine the consistency of responses and the sensitivity to scope.

When conducted according to the exacting standards of the profession, these studies can be very expensive because of the extensive pre-testing and survey work. In addition, while this technique appears easy, its application involves numerous technical challenges. For example, applications of the method are prone to strategic biases on the part of respondents or to structural problems in the design of the questionnaire (Mitchell and Carson, 1989). Question framing, mode of administration, payment formats, and interviewer interactions can all affect the results of contingent market valuation (Cummings *et al.* 1986).

Contingent Valuation Method

The Eagle Mine case study (Kopp and Smith 1992) exemplifies how contingent valuation methods can be applied in resource damage assessments. Contingent valuation questions were presented in both an Eagle County and a statewide survey, to elicit respondents' willingness to pay for the Eagle River cleanup. The Eagle County survey asked respondents about their willingness to make annual payments over 10 years to clean up 200 waste sites involving current legal action. Respondents were given brief descriptions of each site. The survey requested each respondent to perform two allocations: (1) specify from a schedule of percentages the per cent of their total bid for all sites that they would like to assign to the seven sites, and (2) identify a most important site among these seven and the percentage of their bid they would like to have allocated to this one particular site. In addition, respondents were asked to allocate the percentages of their total bid (for cleanup of all 200 sites) that they associated with use and nonuse values. The table below details the results of the analysis. In the Eagle County survey, questions were designed so that the willingness to pay estimates included both use and nonuse values, but allowed for the disaggregation of water and nonwater-based values. In the survey of Colorado residents, no differentiation between water-based and nonwater-based values was possible, but an allocation between use and nonuse values was made. The table displays the mean estimates of annual willingness to pay derived from each survey. In the case of Eagle County residents, the analysts multiplied the annual mean willingness-to-pay estimates by growth of 6063 households, carried forward for 10 years, assuming a population growth of 2 per cent, and then discounted back to 1985 at 10 per cent. The analysts employed a similar aggregation procedure for the statewide estimates.

The quality of a contingent valuation survey questionnaire is sensitive to the amount of information that is known beforehand about the way people think about the underlying natural resource. Certainly, prior information on the ecological attributes or environmental qualities of a particular resource are critical factors in conducting a successful contingent valuation survey. The key point is that, while all the information necessary for assessing an individual's value of the resource is collected in the survey, the analyst must also be able to identify a truly representative sample of well-informed respondents in order to allow extrapolation to the general subject population. Thus, information on who uses the resource and who knows about it is critical.

Eagle Mine Case Contingent Valuation Estimates

Contingent Valuation	Unit Damage Estimate ($1983)	Discounted Present Value of Future Damage (Aggregate Estimate)
Early County willingness-to-pay survey:		
☆ Use and non-use values (water-based)	$73 per year/ household	Country residents $3.4 million
☆ Use and non-use values (non-water-based)	$30-51 per year/ household	Country residents $1.5 million
State of Colorado willingness-to-pay survey:		
☆ Use values	$1.80 per household	$15 million state residents
6,063 household in Eagle Country		
☆ Non-use values	$3.80 per household	$30 million state residents
1.2 million households in Colorado		

Source: Kopp, R. J., and V. K. Smith. 1992. "Eagle Mine and Idarado." In *Natural Resource Damage: Law and Economics*, K. M. Ward and J. W. Duffield (ed.), John Wiley and Sons, Inc. New York, pp. 365-388.

Cross-Cutting Methods

At the present time, there is considerable professional interest in natural resource valuations that are based on cross-cutting methods. These valuation techniques combine elements from market-based methods with pre-existing estimates of natural resource values based on either direct or indirect nonmarket valuation techniques. The interest in applying cross-cutting techniques is motivated by the relative simplicity of using a preexisting study based on an accepted method, as well as the cost considerations in undertaking a fresh natural resource valuation study. Two cross-cutting resource valuation techniques that have gained increased professional attention due to their simplicity and economy of application are discussed here: benefit transfer and unit day value.

Benefit Transfer

Benefit transfer is the use of the estimated values or demand relationship in existing studies to evaluate a site or event for which no site-specific study is available. Given the expense and time associated with the estimation of values of nonmarket natural resources and services, benefit transfer may be a reasonable method by which to determine such values under well-defined conditions. The analyst should consider all available estimates at the onset of the study. Each estimate should be evaluated by comparing the methodology and results of the original studies that may have been undertaken in selecting one that best matches the policy study under consideration. The following criteria have proved to be potentially useful in making this determination:

Benefit Transfer Method

Ulibarri and Ghosh (1995) provide a willingness-to-pay estimate to reduce high particulate matter (PM) levels using the benefit-transfer method. Their application focuses on willingness-to-pay estimates for improved visibility in Benton-Franklin Counties in eastern Washington state. The authors' estimates are based on key parameter values derived by Rowe *et al.* (1980) using a CV survey instrument. In using the Rowe *et al.* parameter estimates, the authors note that their commodity specification (quality of visibility) is similar to the one evaluated by Rowe *et al.* However, to capture the aesthetic realities of the study site, the authors obtained daily observations of PM levels over the period 1990-94 from the Benton-Franklin County Clean Air Authority. In addition, the authors adjusted the various independent variables identified in Rowe *et al.* using county-level census data on the urban/rural population, age distribution, ethnicity and gender, and the levels of household income. Upon making these adjustments, the authors found a measure of the collective willingness to pay across 54,000 household in the Benton-Franklin area of approximately $364,395 per exceedance day, *i.e.*, a day on which PM levels equal 10 or exceed 150 micrograms per meter, the safe minimum standards under the Environmental Protection Agency's National Ambient Air Quality Standards.

☆ Purpose of original value estimates

☆ User group(s) considered

☆ Nature of substitutes in the initial study area

☆ Geographic area

☆ Demographic and socio-economic characteristics

☆ Baseline conditions

☆ Specific or unique problem that may be influenced by the magnitude of the estimates

☆ General attitudes, perceptions, or levels of knowledge

☆ Omitted variables described above.

Once a final set of values has been chosen, consideration should be given to the general magnitudes of the values. If the existing value estimates differ significantly, or if values generated using alternative models differ significantly from one another, consideration should be given to whether they differ in a predictable and consistent manner. In many cases, the defensibility of the transferred economic benefit estimate will depend on the quality of the underlying research. There are no globally accepted, standard criteria by which the quality of existing studies can be judged. Decision-makers should, therefore, seek the guidance of the professional and academic economics community concerning the current minimum conditions for accurate use of the benefit transfer method.

Unit Day Value Method

The unit day value method is similar to the benefit transfer method, except that an average value is derived based on multiple value estimates from existing studies. Consequently, the unit day value of the underlying resource reflects a resource having average preference-related attributes, amenities, or qualities. Any of the valuation approaches described above can potentially serve as underlying studies from which unit day values are drawn. The application of the unit day value method may also involve groups of experts attempting to interpret from the existing set of estimates (regardless of method used in the original study) a best estimate for each of a set of generic types of environmental resources or activities. The unit day value approach then combines and converts these estimates into a standardized unit of measure that reflects the average value of one unit of the resource on a per day basis.

As in many small natural resource damage cases, Loomis and Anderson (1992) relied on existing data and previously estimated equations and values to determine the value of recreational fishing lost as a result of the 1987 spill of Vitavax 200 (a fungicide) in Idaho's Little Salmon River. In attempting to assess damages, Loomis and Anderson found there were no economic valuation studies directly related to the Little Salmon River. To keep assessment costs low, the decision was made not to perform a new study specific to the Little Salmon River but rather to rely on the existing economic survey data. They used a travel-cost demand analysis for several segments of the main Salmon River above and below its confluence with the Little Salmon River, previously undertaken by Donnelly *et al.* (1985). These data had been collected and analyzed as part of an interagency state-federal research effort spanning 1982 to 1985. The survey design and travel cost methodology used in the study followed the spirit of the U.S. Water Resources Council Principles and Standards.

The sample was drawn from a list of individuals who had purchased an Idaho steelhead tag for the 1982 season. The combined mail and telephone survey of 427 anglers had a response rate of 100 per cent. The travel-cost equation and associated values per trip had been peer-reviewed prior to publication as a U.S. Forest Service Experiment Station Bulletin. The available data had been used to estimate a simple quality-augmented zonal travel-cost demand curve. Specifically, one multi-site pooled regional travel-cost demand was estimated for the 11 sections of the Salmon, Clearwater, and Snake Rivers where steelhead fishing was allowed in 1982. Using this demand curve, a value of $25.94 per trip had been calculated for the segment of the Salmon River just downstream from the Little Salmon River. The authors recalculated the per-trip value of $25.94 to a value per steelhead, using information that it took 1.36 trips per steelhead caught. This resulted in a value per steelhead of $35.28. Updating this value from 1982 dollars to 1987 dollars yields $41.52 per fish. With average catch rates, half of the 1688 returning adult steelhead would be recreationally caught. Therefore, the value of the 844 steelhead that were lost due to the spill that would have otherwise been caught was $35,045.

In some cases, unit day values head part-way toward a reasonable benefit transfer approach by developing general categories across activity types or geographical locations. However, unlike benefit transfer, there is no attempt to identify previous studies for comparable sites. For example, the U.S. Water Resources Council guidelines provide unit values across fairly broad activity types and settings. Similarly, the U.S. Forest Service has developed unit values that are specific for activity types and Forest Service regions. However, the analyst must exercise caution when applying such unit day values, insofar as they may reflect a biased selection of studies that reflect poorly on the existing economic value of the natural resource.

Ecological Valuation

The conventional natural resource valuation techniques described above have made little progress in providing a framework to assess the monetary value derived from ecological functions. One reason is that ecological functions are often overlooked in terms of providing preference related value to humans. Thus, the state of the art in natural resource valuation is in search of a framework for addressing natural resource values derived from ecological functions. This section first briefly discusses the emerging field of ecological economics. It then considers *gross primary energy valuation* and *non-glamorous resource valuation*, which are two approaches to measure ecological values in the emerging field of ecological economics.

Ecological Economics

Although controversial, some resource valuation professionals believe that changes in the service flows from ecological systems to human society can be valued in monetary terms, given existing knowledge, scientific data, and estimation techniques. They believe that this would bring such services into management discussions in terms symmetric with marketed goods and services. As a general matter, this could improve the efficiency with which society uses resources. One reason for this view is the belief that such pricing would encourage preservation by making explicit the opportunity cost of development and other economic activities. These people support the continued refinement and extensions of economic valuation techniques based on people's preferences over ecological resources.

Other experts express serious reservations about the prospects of deriving willingness-to-pay estimates for ecological resources. Their distrust arises in part because of the potential lack of knowledge associated with people's understanding of ecological functions and how ecological systems are damaged through human activities. Without a firm understanding of the ecological impacts of human intervention, there is no reliable way to estimate meaningful ecological damage. Accordingly, there is skepticism as to whether monetary values can be assigned to damages that might arise within the intricate web of ecological interdependencies in both small- and large-scale ecosystems. Nevertheless, monetary values are beginning to surface in the ecological economics literature. One approach is based on the energy valuation of gross primary production, which incorporates both economic and ecological values in one index.

Gross Primary Energy Valuation

This procedure has been applied to the valuation of different wetland types (Constanza *et al.* 1989). It is argued that estimates of gross primary production have merit since the entire food chain depends upon this primary production. The methodology is not without its problems, however. For instance, it is not well understood whether those species supported by a particular food chain have equal social values. In general, the embodied energy approach measures only ecologically based values. Unlike an economic valuation approach, values for such functions and services as storm protection, aesthetics, and water treatment are completely ignored.

Recently, an alternative cross-cutting approach to environmental issues has come under the rubric of ecological-economic valuation. A team of ecologists and economists sponsored by the U.S. Environmental Protection Agency (EPA) has undertaken the development of a crosscutting model of an ecological-economic system (*i.e.*, the Maryland Patuxent Watershed), with emphasis on wetlands. Preliminary work is based on a coastal ecological landscape spatial simulation model developed by Robert Costanza and colleagues from the University of Maryland. Ultimately, the cross-cutting model may incorporate economic behavior, thereby capturing interrelationships between human activities, the ecosystem, and ecological valuation. A related attempt at employing cross-cutting techniques in determining ecological values is based on the recognition that humans may have preferences relating to the functions and services performed by ecological resources, referred to as ecological resource valuation.

Ecological Resource Valuation

The need for a framework addressing the value of ecological functions is particularly acute in assessing policy choices that affect the integrity of ecological systems. Using the example of wind-blown dust, Ulibarri and Ghosh (1995) suggest that these policy decisions require a weighting of ecological values based on two related subsets of information: what is valued by humans as an eco-good (*i.e.*, cleaner air) and what has intrinsic value to the natural eco-system (*i.e.*, vegetative cover). Using the term ecological resources, the authors focused attention on resource services that are functionally important to ecosystems but frequently overlooked in terms of providing value to humans. Such resources have received very little attention relative to their more glamorous cousins, such as endangered salmon runs or old-growth timber stands.

The authors note that the key objectives of ecological resource valuation are (*a*) to provide a framework that aggregates the values of goods and services rendered by selected ecological functions and (*b*) to determine defensible upper and lower limits on these values. The possibility of interpolating between these limits would enable a more robust estimation of the value of eco-goods and services, allowing policy makers to form a more complete understanding of the benefits and costs of ecological preservation.

The preliminary work undertaken by Scott *et al.* (1997) considered social values associated with undeveloped shrub-steppe sites; these are arid environs which are traditionally overlooked in land-use decisions. Relative to the perceived values, the authors attempted applications of the benefit transfer method, the travel cost method, and the method of hedonic damage-pricing. In order to estimate the intrinsic values of natural ecosystems, they applied a replacement cost methodology based on the idea of replacing the functions performed by the natural ecosystem through a human engineered analog. Using these cross-cutting resource valuation techniques, the authors maintained that the economic value of shrub-steppe sites reflects both their ecological services and recreational uses. Given the uncertainty that exists as to the social benefits from preserving undeveloped shrub-steppe, they suggest the need for further analysis in order to establish credibility in ecological site valuations.

Unresolved Issues

The following discussion focuses on a series of distinct and challenging issues in the valuation of natural resources and the environment: (*a*) the choice of a *discount rate* in assessing the present and future values of benefits and costs; (*b*) the individual's *time-preference* in deriving benefits over the near term as opposed to later on; (*c*) the role of *equity and fairness* in resource valuations involving present and future generations; (*d*) the conceptual understanding of *risk and uncertainty* in the valuation of natural resources and the environment; and (*e*) qualification of the *measurement errors* in the application of the natural resource valuation techniques discussed in this handbook. Without identifying the potential importance of these factors in the valuation process, the analysis of natural resource values will remain incomplete and contentious if used by DOE field operations as a basis for decision-making.

Discounting and Time Preference

Discount rates enable one to determine the present value of the benefits and costs associated with the future use and enjoyment of natural resources. If the analyst of future benefits and costs sets a high discount rate, say 10 per cent, the present value of benefits in the distant future becomes insignificant when compared with the present value of benefits in the near-term future. For example, using continuous discounting at the rate of 10 per cent, the present value of $1000 of benefits obtained 2 years in the future is $818.73, and in 10 years becomes $367.88, less than half as much. By choosing a lower discount rate, say 2 per cent, the analyst reduces this temporal bias: $1000 of benefits 10 years hence becomes $818.73. Given the implications of this basic arithmetic on natural resource valuations, it is no wonder there is so much controversy among economists, scientists, and policy makers over the applications of appropriate discount rates.

Despite the controversies that exist, discounting is part of applied benefit-cost analyses. Its main role is helping to evaluate a series of costs or benefits that are strung out over the future. Discounting is a way of adding up a series of future net benefits into an estimate of present value. However, as we have already seen, the outcome of this exercise depends on which particular discount rate we use. Under rates of 2 per cent, we are essentially treating a dollar of benefits from a natural resource in one year as very similar in value to a dollar of benefits in any other year. Using very high rates, say above 7 per cent, we are saying that a dollar of natural resource benefits in the near-term is much more valuable to us than it would be later on. Thus, the higher the discount rate used in the calculation of present values, the more we are favoring the near term use and/or enjoyment of natural resources relative to more distant future uses. To the contrary, the lower the discount rate used in present value calculations, the more equally we are weighing the benefits over time.

Chapter 8

Dose-Response Modelling

The biological basis for dose-response models derives from major steps in the disease process as they result from the interactions between the pathogen, the host and the matrix. Figure 1 illustrates the major steps in the overall process, with each step being composed of many biological events. Infection and illness can be seen as resulting from the pathogen successfully passing multiple barriers in the host. These barriers are not all equally effective in eliminating or inactivating pathogens and may have a range of effects, depending on the pathogen and the individual. Each individual pathogen has some particular probability to overcome a barrier, which is conditional on the previous step(s) being completed successfully. The disease process as a whole and each of the component steps may vary by pathogen and by host. Pathogens and hosts can be grouped with regard to one or more components, but this should be done cautiously and transparently.

A dose-response model describes the probability of a specified response from exposure to a specified pathogen in a specified population, as a function of the dose.

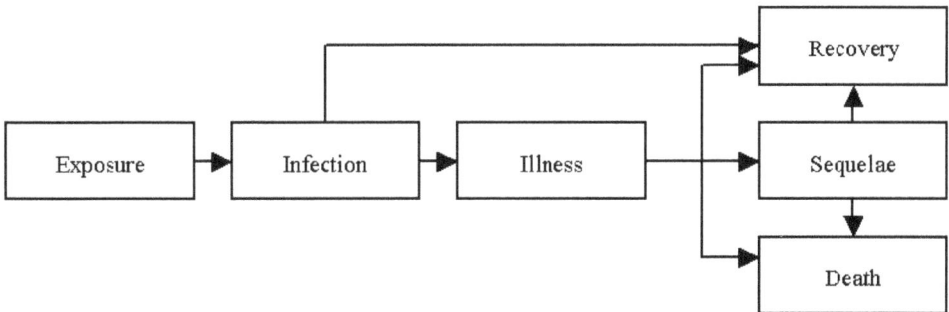

Figure 1: The major steps in the foodborne infectious disease process.

This function is based on empirical data, and will usually be given in the form of a mathematical relationship. The use of mathematical models is needed because:

☆ Contamination of food and water usually occurs with low numbers or under exceptional circumstances; the occurrence of effects cannot usually be measured by observational methods in the dose range needed, and hence models are needed to extrapolate from high doses or frequent events to actual exposure situations;

☆ Pathogens in food and water are usually not randomly dispersed but appear in distinct clumps or clusters, which must be taken into account when estimating health risks; and

☆ Experimental group sizes are limited, and models are needed, even in well controlled experiments, to distinguish random variation from true biological effects.

Plots of empirical datasets relating the response of a group of exposed individuals to the dose (often expressed as a logarithm) frequently show a sigmoid shape, and can be fitted by a large number of mathematical functions. However, when extrapolating outside the region of observed data, these models may predict widely differing results (cf. Coleman and Marks, 1998; Holcomb *et al.*, 1999). It is therefore necessary to select between the many possible dose-response functions. In setting out to generate a dose-response model, the biological aspects of the pathogen-host-matrix interaction should be considered carefully. The model functions derived from this conceptual information should then be treated as *a priori* information.

Exposure

In general, biologically plausible dose-response models for microbial pathogens should consider the discrete (particulate) nature of organisms and should be based on the concept of infection from one or more "survivors" from an initial dose. Before proceeding, however, it is necessary to carefully consider the concept of "dose".

The concentration of pathogens in the inoculum is usually analysed by some microbiological, biochemical, chemical or physical method. Ideally, such methods would have 100 per cent sensitivity and specificity for the target organism, but this is rarely the case. Therefore it may be necessary to correct the measured concentration for the sensitivity and specificity of the measurement method to provide a realistic estimate of the number of viable, infectious agents. The result may be greater or smaller than the measured concentration. Note that, in general, the measurement methods used to characterize the inoculum in a data set used for dose-response modelling will differ from the methods used to characterize exposure in a risk assessment model. These differences need to be accounted for in the risk assessment.

Multiplying the concentration of pathogens in the inoculum by the volume ingested, the mean number of pathogens ingested by a large group of individuals can be calculated. The actual number ingested by any exposed individual is not equal to this mean, but is a variable number that can be characterized by a probability distribution. It is commonly assumed that the pathogens are randomly distributed in the inoculum, but this is rarely the case. Compound distribution (or over-dispersion) may result from two different mechanisms:

☆ A 'unit" as detected by the measurement process (*e.g.*, a colony-forming unit (CFU), a tissue culture infectious dose, or a Polymerase Chain Reaction (PCR) detectable unit) may, due to aggregation, consist of more than one viable, infectious particle. This is commonly observed for viruses, but may also be the case for other pathogens. The degree of clumping strongly depends on the methods used for preparing the inoculum.

☆ In a well-homogenized liquid suspension, unit doses will be more or less randomly distributed. If the inoculum consists of a solid or semisolid food matrix, however, spatial clustering may occur and result in over-dispersion of the inoculum. This aspect may differ between the data underlying the dose-response model and the actual exposure scenario.

The Poisson distribution is generally used to characterize the variability of the individual doses when pathogens are randomly distributed. Micro-organisms have a tendency to aggregate in aqueous suspensions. In such cases, the number of "units" counted is not equal to the number of infectious particles but to the number of aggregates containing one or more infectious particles. In such cases, it is important to know whether the aggregates remain intact during inoculum preparation or in the gastrointestinal tract. Also, different levels of aggregation in experimental samples and in actual water or food products need to be accounted for.

Infection

Each individual organism in the ingested dose is assumed to have a distinct probability of surviving all barriers to reach a target site for colonization. The relation between the actual number of surviving organisms (the effective dose) and the probability of colonization of the host is a key concept in the derivation of dose-response models, as will be discussed later.

Infection is most commonly defined as a situation in which the pathogen, after ingestion and surviving all barriers, actively grows at its target site (Last, 1995). Infection may be measured by different methods, such as faecal excretion or immunological response. Apparent infection rates may differ from actual infection rates, depending on the sensitivity and specificity of the diagnostic assays. Infection is usually measured as a quantal response (presence or absence of infection by some criterion). The use of continuous-response variables (*e.g.*, an antibody titre) may be useful for further development of dose-response models. Infections may be asymptomatic, where the host does not develop any adverse reactions to the infection, and clears the pathogens within a limited period of time, but infection may also lead to symptomatic illness.

Illness

Microbial pathogens have a wide range of virulence factors, and may elicit a wide spectrum of adverse responses, which may be acute, chronic or intermittent. In general, disease symptoms may result from either the action of toxins or damage to the host tissue. Toxins may have been preformed in the food or water matrix ("intoxication") or may be produced *in vivo* by microorganisms in the gut ("toxico-infection"), and may operate by different pathogenic mechanisms (*e.g.*, Granum, Tomas and Alouf, 1995). Tissue damage may also result from a wide range of mechanisms, including destruction of host cells, invasion and inflammatory responses. For many food borne pathogens, the precise pathogenic sequence of events is unknown, and is likely to be complex. Note that health risks of toxins in water (*e.g.*, cyan bacterial toxins) usually relate to repeated exposures, and these require another approach, which resembles hazard characterization of chemicals.

Illness can basically be considered as a process of cumulative damage to the host, leading to adverse reactions. There are usually many different and simultaneous signs and symptoms of illness in any individual, and the severity of symptoms varies among pathogens and among hosts infected with the same pathogen. Illness is therefore a process that is best measured on a multidimensional, quantitative, continuous scale (number of stools passed per day, body temperature, laboratory measurements, etc.). In contrast, in risk assessment studies, illness is usually interpreted as a quantal response (presence or absence of illness), implying that the results depend strongly on the case definition. A wide variety of case definitions for gastrointestinal illness are used in the literature, based on a variable list of symptoms, with or without a specified time window, and sometimes including laboratory confirmation of etiological agents. This lack of standardization severely hampers integration of data from different sources.

Sequelae and Mortality

In a small fraction of ill persons, chronic infection or sequelae may occur. Some pathogens, such as *Salmonella enterica* serotype Typhi, are invasive and may cause bacteraemia and generalized infections. Other pathogens produce toxins that may result not only in enteric disease but also in severe damage in susceptible organs. An example is haemolytic uraemic syndrome, caused by damage to the kidneys from Shiga-like toxins of some *Escherichia coli* strains. Complications may also arise by immune-mediated reactions: the immune response to the pathogen is then also directed against the host tissues. Reactive arthritis (including Reiter's syndrome) and Guillain-Barré syndrome are well known examples of such diseases. The complications from gastroenteritis normally require medical care, and frequently result in hospitalization. There may be a substantial risk of mortality in relation to sequelae, and not all patients may recover fully, but may suffer from residual symptoms, which may last a lifetime. Therefore, despite the low probability of complications, the public health burden may be significant. Also, there is a direct risk of mortality related to acute disease, in particular in the elderly, neonates and severely immuno-compromised.

Modelling Concepts

Several key concepts are required for the formulation of biologically plausible dose-response models. These relate to:

☆ Threshold vs non-threshold mechanisms;

☆ Independent vs synergistic action; and

☆ The particulate nature of the inoculum.

Each of these concepts will be discussed below in relation to the different stages of the infection and disease process. Ideally, the dose-response models should represent the following series of conditional events: the probability of infection given exposure; the probability of acute illness given infection; and the probability of sequelae or mortality given acute illness.

In reality, however, the necessary data and concepts are not yet available for this approach. Therefore models are also discussed that directly quantify the probability of illness or mortality given exposure.

Threshold vs Non-Threshold Mechanisms

The traditional interpretation of dose-response information was to assume the existence of a threshold level of pathogens that must be ingested in order for the microorganism to produce infection or disease. A threshold exists if there is no effect below some exposure level, but above that level the effect is certain to occur. Attempts to define the numerical value of such thresholds in test populations have typically been unsuccessful, although the concept is widely referred to in the literature as the "minimal infectious dose".

An alternative hypothesis is that, due to the potential for micro-organisms to multiply within the host, infection may result from the survival of a single, viable, infectious pathogenic organism ("single-hit concept"). This implies that, no matter how low the dose, there is always, at least in a mathematical sense, and possibly very small, a non-zero probability of infection and illness. Obviously, this probability increases with the dose.

Note that the existence or absence of a threshold, at both the individual and population levels, cannot be demonstrated experimentally. Experimental data are always subject to an observational threshold (the experimental detection limit): an infinitely small response cannot be observed. Therefore, the question of whether a minimal infectious dose truly exists or merely results from the limitations of the data tends to be academic. A practical solution is to fit dose-response models that have no threshold (no mathematical discontinuity), but are flexible enough to allow for strong curvature at low doses so as to mimic a threshold-like dose-response.

The probability of illness given infection depends on the degree of host damage that results in the development of clinical symptoms. For such mechanisms, it seems to be reasonable to assume that the pathogens that have developed *in vivo* must exceed a certain minimum number. A non-linear relation may be enforced because the interaction between pathogens may depend on their numbers *in vivo*, and high numbers are required to switch on virulence genes (*e.g.*, density dependent quorum-sensing effects). This concept, however, is distinct from a threshold for administered dose, because of the possibility, however small, that a single ingested organism may survive the multiple barriers in the gut to become established and reproduce.

Independent Action vs Synergistic Action

The hypothesis of independent action postulates that the mean probability p per inoculated pathogen to cause (or help cause) an infection (symptomatic or fatal) is independent of the number of pathogens inoculated, and for a partially resistant host it is less than unity. In contrast, the hypotheses of maximum and of partial synergism postulate that inoculated pathogens cooperate so that the value of p increases as the size of the dose increases (Meynell and Stocker, 1957). Several experimental studies have attempted to test these hypotheses and the results have generally been consistent with the hypothesis of independent action (for a review, see Rubin, 1987).

Quorum sensing is a new area of research that is clearly of importance in relation to the virulence of some bacteria. It means that some phenotypic characteristics such as specific virulence genes are not expressed constitutively, but are rather cell-density dependent, using a variety of small molecules for cell-to-cell signalling, and are only expressed once a bacterial population has reached a certain density (De Kievit and Iglewski, 2000). While the biology of quorum sensing and response is still being explored, the nature of the effect is clear, it may be that some virulence factors are only expressed once the bacterial population reaches a certain size. The role of quorum sensing in the early stages of the infectious process has not been investigated in detail, and no conclusion can be drawn about the significance of quorum sensing in relation to the hypothesis of independent action. In particular, the role of interspecies and intraspecies communication is an important aspect. Sperandio *et al.* (1999) have demonstrated that intestinal colonization by enteropathogenic *E. coli* could be induced by quorum sensing of signals produced by non-pathogenic *E. coli* of the normal intestinal flora.

Selection of Models

Specific properties in the data become meaningful only within the context of a model. Different models may, however, lead to different interpretations of the same data, and so a rational basis for model selection is needed. Different criteria may be applied when selecting mathematical models. For any model to be acceptable, it should satisfy the statistical criteria for goodness of fit. However, many different models will usually fit a given data set (for example, see Holcomb *et al.*, 1999) and therefore goodness of fit is not a sufficient criterion for model selection. Additional criteria that might be used are conservativeness and flexibility.

Conservativeness can be approached in many different ways: "Is the model structure conservative?" "Are parameter estimates conservative?" "Are specific properties of the model conservative?" and so forth. It is not recommended to build conservativeness into the model structure itself. From a risk assessment perspective, a model should be restricted to describing the data and trying to discriminate the biological signal from the noise. Adding parameters usually improves the goodness of fit of a model, but using a flexible model with many parameters may result in greater uncertainty of estimates, especially for extrapolated doses. Flexible models and sparse datasets may lead to overestimation of the uncertainty, while a model based on strong assumptions may be too restrictive and lead to underestimation of the uncertainty in risk estimates.

It is recommended that dose-response models be developed based on a set of biologically plausible, mechanistic assumptions, and then to perform statistical analysis with those models that are considered plausible. Note that it is generally not possible to "work back", *i.e.*, to deduce the assumptions underlying a given model formula. There is a problem of identifiability: the same functional form may result from different assumptions, while two (or more) different functional forms (based on different assumptions) may describe the same dose-response data equally well. This may result either in very different fitted curves if the data contains little information, or virtually the same curves if the data contain strong information. However, even in the last case, the model extrapolation may be very different. This means that a choice between different models or assumptions cannot be made on the basis of data alone.

Dose-Infection Models

The foregoing considerations lead us to the working hypothesis that, for microbial pathogens, dose-infection models based on the concepts of single-hit and independent action are regarded as scientifically most plausible and defensible. When the discrete nature of pathogens is also taken into account.

The single-hit models are a specific set of models in a broader class of mechanistic models. Haas, Rose and Gerba (1999) describe models that assume the existence of thresholds – whether constant or variable – for infection, *i.e.*, some minimum number of surviving organisms larger than 1 is required for the infection to occur. Empirical (or tolerance distribution) models, such as the log-logistic, log-probit and Weibull (-Gamma) models, have also been proposed for dose-response modelling. The use of these alternative models is often motivated by the intuitive argument that single-hit models overestimate risks at low doses.

Infection-Illness Models

Currently, infection-illness models have received little attention and data available are extremely limited. Experimental observations show that the probability of acute illness among infected subjects may increase with ingested dose, but a decrease has also been found (Teunis, Nagelkerke and Haas, 1999), and often the data do not allow conclusions about dose dependence, because of the small numbers involved. Given this situation, constant probability (*i.e.*, independent of the ingested dose) models, possibly stratified for subgroups in the population with different susceptibilities, seem to be a reasonable default. Together with ingested dose, illness models should take into account the information available on incubation times, duration of illness and timing of immune response, and should preferably measure illness as a multidimensional concept on continuous scales. There is no basis yet to model the probability of illness as a function of the numbers of pathogens that have developed in the host.

Dose-Illness Models

The default assumption of constant probability models for illness given infection lead to the conclusion that the only difference between dose-infection and dose-illness models is that the dose-illness models do not need to reach an asymptote of 1, but of P(ill | inf). They would essentially still belong to the family of hit-theory models.

Sequelae and Mortality

Given illness, the probability of sequelae or mortality, or both, depends of course on the characteristics of the pathogen, but more importantly on the characteristics of the host. Sequelae or mortality are usually rare events that affect specific subpopulations. These may be identified by factors such as age or immune status, but increasingly genetic factors are being recognized as important determinants. As above, the current possibilities are mainly restricted to constant probability models. Stratification appears to be necessary in almost all cases where an acceptable description of risk grouping is available.

Extrapolation

Low Dose Extrapolation

Dose-response information is usually obtained in the range where the probability of observable effects is relatively high. In experimental studies using human or animal subjects, this is related to financial, ethical and logistical restrictions on group size. In observational studies, such as outbreak studies, low dose effects can potentially be observed directly, but in these studies only major effects can be distinguished from background variation. Because risk assessment models often include scenarios with low dose exposures, it is usually necessary to extrapolate beyond the range of observed data. Mathematical models are indispensable tools for such extrapolations, and many different functional forms have been applied. Selection of models for extrapolation should primarily be driven by biological considerations, and only subsequently by the available data and their quality. The working hypotheses of no-threshold and independent action lead to a family of models that is characterized by linear low dose extrapolations on the log/log scale, or even on the arithmetic scale. That is, in the low dose range, the probability of infection or disease increases linearly with the dose. On the log-scale, these models have a slope of 1 at low doses. Some examples include:

- ☆ The exponential model $P = r.D$
- ☆ Beta-Poisson model $P = (a/b).D$
- ☆ The hypergeometric model $P = \{a/(a+b)\}.D$

where, D = mean ingested dose and r, a and b are model parameters. Note that if $a > b$, the risk of infection predicted by the Beta-Poisson model is larger than the risk of ingestion, which is not biologically plausible. This highlights the need to carefully evaluate the appropriateness of using this simplified model for analysing dose-response data.

Extrapolation in the Pathogen-Host-Matrix Triangle

Experimental datasets are usually obtained under carefully controlled conditions, and the data apply to a specific combination of pathogen, host and matrix. In actual exposure situations, there is more variability in each of these factors, and dose-response models need to be generalized. Assessing such variability requires the use of multiple datasets that capture the diversity of human populations, pathogen strains and matrices. Failure to take such variation into account may lead to underestimation of the actual uncertainty of risks.

When developing dose-response models from multiple datasets, one should use all of the data that is pertinent. There is currently no way of determining which data source is best. This requires that the risk assessor make choices. Such choices should be based on objective scientific arguments to the maximum possible extent, but will inevitably include subjective arguments. Such arguments should be discussed with the risk manager and their significance and impact for risk management considered. The credibility of dose-response models increases significantly if dose-response relations derived from different data sources are consistent, especially if the data are of varying types.

When combining data from different sources, a common scale on both axes is needed. This often requires adjusting the reported data to make them comparable. For dose, test sensitivity, test specificity, sample size, etc., need to be taken into account. For response, a consistent case definition is needed or the reported response needs to be adjusted to a common denominator (*e.g.*, infection conditional probability of illness given infection). Combining data from different sources within a single (multilevel) dose-response model requires thorough statistical skills and detailed insight into the biological processes that generated the data. An example is the multilevel dose-response model that has been developed for different isolates of *Cryptosporidum parvum* (Teunis *et al.*, 2002a). The issue of combining data from different outbreak studies is discussed in the FAO/WHO risk assessments of *Salmonella* in eggs and broiler chickens (FAO/WHO, 2002a).

Dose-response relations where an agent only affects a portion of the population may require that subpopulation to be separated from the general population in order to generate meaningful results. Using such stratified dose-response models in actual risk assessment studies requires that the percentage of the population that is actually susceptible can be estimated. Consideration of such subpopulations appears to be particularly important when attempting to develop dose-response relations for serious infections or mortality. However, it would also be pertinent when considering an agent for which only a portion of the population can become infected.

Stratified analysis can also be useful when dealing with seemingly outlying results, which may actually indicate a subpopulation with a different response. Removal of one or more outliers corresponds to removing (or separately analysing) the complete group from which the outlying results originated. Where a specific reason for the separation cannot be identified, there should be a bias toward being inclusive in relation to the data considered. Any elimination of the data should be clearly communicated to ensure the transparency of the assessment.

A particular and highly relevant aspect of microbial dose-response models is the development of specific immunity in the host. Most volunteer experiments have been conducted with test subjects selected for absence of any previous contact with the pathogen, usually demonstrated by absence of specific antibodies. The actual population exposed to foodborne and waterborne pathogens will usually be a mixture of totally naive persons and persons with varying degrees of protective immunity. No general statements can be made on the impact of these factors. This is strongly dependent on the pathogen and the host population. Some pathogens, such as many childhood diseases and the hepatitis A virus, will confer lifelong immunity upon first infection whether clinical or subclinical, whereas immunity to other pathogens may wane within a few months to a few years, or may be evaded by antigenic drift. At the same time, exposure to non-pathogenic strains may also protect against virulent variants. This principle is the basis for vaccination, but has also been demonstrated for natural exposure, *e.g.*, to non-pathogenic strains of *Listeria monocytogenes* (Notermans *et al.*, 1998). The degree to which the population is protected by immunity depends to a large extent on the general hygienic situation. In many developing countries, large parts of the population have built up high levels of immunity, and this is thought to be responsible for lower incidence or less serious forms of illness. Some examples are the predominantly watery form of diarrhoea by *Campylobacter* spp. infections in children and the lack of illness from this organism in young adults in developing countries. The apparent lack of *E. coli* O157:H7-related illness in Mexico has been explained as the result of cross-immunity following infections with other *E. coli*, such as enteropathogenic *E. coli* strains that are common there. In contrast, in the industrialized world, contact with enter pathogens is less frequent and a larger part of the population is susceptible. Obviously, age is an important factor in this respect.

Chapter 9
Cost–Benefit Analysis

Economics can be derived into two areas of positive and normative economics. Positive economics attempts to explain and predict actual economic activity, aiming at the representation of the 'facts' without judging their contribution to the economic development. In contrast, normative economics explicitly introduces value judgments focusing on the efficiency evaluation of each investment project. Economic appraisal techniques aim to indicate policy makers which options are 'better' and such a process falls firmly in normative economics. Cost-benefit analysis provides the answer to the normative approach for the definition of economic impacts related to environmental investments and policies.

In a close and detailed examination of cost-benefit analysis we investigate whether an investment project or an environmental policy, contributes to social prosperity. An action or policy is considered solvent if the net social benefits counterbalance the respective net social cost.

Environmental Impact Evaluation Techniques

Extracting reliable estimations of environmental costs caused by human activities, more often than not, constitutes a major drawback in valuating the net social cost. In the case of 'tradeables' the real economic values are assessed by means of willingness to pay for purchasing them. On the contrary, whenever the goods and services in question are 'non-tradeables' then the usage of alternative cost estimation methods is required. Three general evaluation approaches of nontradeable goods exist, each one of these comprise a series of diverse methods:

Revealed preference techniques (direct valuation), which derive from preferences from actual, observed, market-based information. These preferences for environmental goods and services are revealed indirectly when individuals purchase marketed goods which are related to the environmental good some way. More specifically, the essence of these techniques is that they infer environmental values from markets in which environmental factors have an immediate influence. For example, these are markets for certain goods to which environmental commodities are related, as either substitutes or complements to the goods in question. In this way people's actions in actual markets reflect, to a certain extent, their preferences for environmental assets. There are three main revealed preference techniques (*a*) hedonic pricing, (*b*) travel cost method, and (*c*) replacement cost.

Stated preference techniques (indirect valuation), which attempt to elicit preferences in direct way by use of questionnaire. The same techniques enable economic values to be estimated for a wide range of commodities which are not traded in real markets. Additionally, these techniques are the only way to estimate the non-use value of environmental resources. In this study two approaches of this kind are examined (*a*) contingent valuation, and (*b*) conjoint analysis.

Production function approaches, which link environmental quality changes to changes in production relationships. These approaches are indirect means of non-market good evaluation, related either to firms productions goods or services, or to household producing services that generate positive utility. They also share the characteristic that changes in expenditures represent the need to substitute other inputs for changes in environmental quality. We consider two such approaches (*a*) averting behaviour, and (*b*) dose-response functions. Linking the two is the evaluation of human morbidity and mortality, a crucial issue which often determines project approval and environmental standards.

We consider each of these approaches in turn, highlighting when each could be used, their advantages and drawbacks as well as their applicability to potential environmental impact receptors.

Production Function Approaches Averting Behaviour

The basis for the averting behaviour technique is the observation that marketed goods can act as substitutes for environmental goods in certain circumstances. When a decline in environmental quality occurs, expenditures can be made to mitigate the effects and protect the household from welfare reductions. The method is applicable in situations where households spend money to offset environmental impacts. It requires data on the environmental change and its associated

substitution effects. Fairly crude approximations can be found by simply looking directly at changes in expenditures on the substitute good resulting from some environmental change. The main advantages of this method is the relatively modest data requirements and the fact that provides estimations based on actual expenditures. Simultaneously, it is often observed beneficial effects not to be concerned explicitly resulting in incorrect estimates. Furthermore, in some cases important aspects of individuals' behavioural responses are ignored or underestimated increasing the possibility of false estimations.

Price of Marketed Natural Resources

Several types of natural resources are traded and have a market price: *e.g.*, water for drinking and irrigation, timber, fish, and iron ore. In project appraisal, the question of how many people actually pay for such resources as assessed in financial analysis is becoming extremely important, because market prices of environmental resources affect people's behaviour, and their use of such resources. The lower the price is, the greater the incentive will be to exploit natural resources. Higher prices encourage conservation.

Consider a producer of textiles, who pollutes a river adjacent to the factory. If the polluter does not pay, the environmental damage is not expressed in production costs and sales prices. The relatively low sale price provides an incentive to consumers to buy textiles, and (indirectly) contribute to environmental decay. If the factory owner is obliged to pay for pollution (environmental costs are internalised), he is faced with pollution charges or investment in measures to prevent pollution. Either way, production costs will increase and so probably will consumer prices. A lower demand for the final product will lead to less production, which leads to less pollution.

Irrigation water is often heavily subsidised or even provided free of charge. This is harmful to the environment. Firstly, low prices encourage water use and wastage, which may lead to lower groundwater levels. Secondly, because farmers pay very little for water, organisations responsible for maintenance have limited financial resources. Inadequate maintenance may lead to environmental problems, such as siltation and erosion.

The examples show that marketed natural resources raise important questions in both financial and economic analysis. Financial analysis focuses on the price people pay, whether they are free market prices or user charges determined by the government. In the economic analysis the appropriateness of this market price is investigated. To what extent is the market price a guide to the value of environmental goods to society? The *economic price* of a natural resource, for instance timber, may have several components:

☆ *Direct costs,* such as clearing land, felling trees, and distribution, which are usually the only costs included in domestic and world market prices (and sometimes only partly so);

☆ The cost of *externalities*: deforestation may lead to erosion and loss of farm income, loss of income opportunities of tribes, loss of species, etc.

☆ *User costs* or future benefits foregone: a resource used now is not available for future use and income generation;

☆ *Existence value,* that is the appreciation of a natural resource not used now or in the future.

It is not always possible to estimate the full economic cost of a natural resource. Nevertheless, a comparison between an *incomplete economic* price and the actual market price may already lead to interesting conclusions. If the market price is lower than the incomplete economic price (for instance only the production function is being taken into account), then market prices are certainly too low.

Incorporation of Non-Marketed Environmental Effects

Similar to other resources, the use of environmental resources should be accounted for. Environmental effects should be estimated both on-site and off-site.

For instance, in the case of soil erosion, on-site impacts may be reduced soil depth and loss of nutrients. Off-site effects may include increased sedimentation in downstream hydro-electric schemes and reservoirs, impacts on aquatic life, and flooding due to increased run-off.

Measurement of these impacts may cause problems because of limited knowledge of ecosystems and of how human activities affect the environment. The collection of reliable information about environmental effects may be time-consuming and costly.

Environmental impacts should be estimated first in physical terms. For example, the percentage increase of emissions of gases, area of land lost due to erosion, and temperature increase in degrees Celsius. In the next step they need to be monetized, that is given a *monetary value*.

Environmental effects are often external effects, which by definition do not come into a financial CBA analysis. However, in projects aimed at environmental protection or improvement, financial and ecological variables may be related.

Valuation of Effects

Environmental amenities, such as clean air, species, natural beauty and the ozone layer, cannot be traded because no one is the (sole) owner. Thus there is no direct price indicator for their value to society. Even with natural resources that are traded (timber, land, fish), the market price is usually a poor measure of the value to society.

Fortunately, economists are not without tools to value the environment. To understand the principle underlying these methods, consider the question of how to value erosion. Assume that erosion reduces agricultural output. The value of the ecological effect (increased erosion) may be estimated by investigating the loss of agricultural production it causes. Similarly, the ecological benefits of an erosion control project (reduced erosion) may be valued in terms of the resulting increase (or more precise by averted decrease) in agricultural output. Consequently, the value of ecological costs is *benefits foregone* in agriculture; a proxy for the value of ecological benefits is the *avoided costs* in agriculture.

A basic principle of valuation techniques is that environmental effects that do not have market value or have an artificially low market value are linked to markets, where prices are available. These may be *conventional* markets, *surrogate* markets, or *hypothetical* markets. In a conventional market, the value of a productive natural resource may be associated with the value of the final output. In surrogate markets, the valuation focuses on the contribution of the factor of environmental quality to the value of man-made goods and services. Artificial markets are simulated by asking people to value environmental goods and services.

Valuation techniques are classified into two groups:

- ☆ *Cost approaches*: actual or hypothetical expenditures aimed at influencing environmental quality;
- ☆ *Benefit approaches*: the impacts of changes in environmental quality on income opportunities.

Cost Approaches

Individuals and organizations may invest in facilities aimed at preventing or mitigating environmental problems. Preventive measures are source- or process-oriented, whereas mitigating measures are effect-oriented measures to combat occurring damage. Both types of measures imply *defensive expenditures* that may be taken as a minimum value for environmental quality changes.

Changes in environmental quality may cause damage to physical assets, such as houses, buildings, machines and cars. The value of environmental changes may be estimated by investigating the costs of *replacing* lost assets or *relocating* these assets.

Due to development projects, irreversible damage may be inflicted on environmental goods and services. A *shadow project* may be implemented to create as much environmental benefit as that lost due to the original project.

The valuation techniques illustrated above all related environmental quality changes to existing markets for man-made products. If such markets do not exist, a hypothetical market may be created by asking individuals how they value environmental services. This is the so-called *contingent valuation method*.

In surveys, people may express their willingness-to-pay for being protected against environmental degradation, for clean water, nature and so on. Their answers are contingent to the hypothetical market that has been described to them. Alternatively, they may express their willingness-to-accept financial compensation for being exposed to environmental decay.

Benefit Approaches

Environmental services are often inputs to the production of marketed goods and services. Consequently, a change in environmental quality will directly affect *productivity* levels in incomegenerating activities. A lower productivity implies that from a given quantity of resources less production can be derived, and consequently less income. The value of a loss in environmental quality can be expressed in terms of foregone income.

Changes in environmental quality may affect human health and subsequently human productivity. The *loss of earnings* approach (also know as human capital or foregone earnings approach) estimates environmental value by calculating the loss of income and costs of medical treatment that may be attributed to environmental problems.

Lack of sewerage facilities may lead to pollution of drinking water. As a result, people may become ill more often, and families experience a loss in income and face higher outlays for doctor visits and medicines

Usually it is not possible to make an integral assessment of the value of lost ecosystems because they have multiple functions about which we have limited knowledge. Rather than to attempt to assign a value to each function separately, an economist may raise the question of how much income would be lost if the ecosystem was preserved. This refers to the *opportunity costs* of preservation: benefits foregone if natural resources are not degraded.

This may be illustrated by the example of the construction of a dam for the generation of hydroelectric power, for which a unique wilderness area would have to be sacrificed. The loss of this ecosystem should be included in economic CBA. Its value may be estimated in terms of the costs to the nation if the dam would not be built in that area. Alternative and more expensive power generation options would be required. The value of the ecosystem, or the opportunity costs of preservation, may be based on the additional costs of the alternative projects

Cost Benefit Calculations, Assessments and Analyses Mean different Things

Infrastructural investments and other measures in the transport sector can lead to different kinds of economic effects. Examples of such effects are changed travel times, accident risks, emissions and encroachment on natural and cultural environments. A *cost benefit calculation* is a calculation that includes the effects that could be identified, quantified and valued in monetary terms. Net present value ratio calculation is an example of this type of calculation. The step from cost benefit calculation to *cost benefit assessment* is to include the effects that have been identified as relevant but could not be quantified or valued in monetary terms. The concept *cost benefit analysis* is an overall concept for all analyses of a cost benefit nature which can be made of alternative courses of action.

Calculation Values are Adjusted Upwards to take into Consideration Price Increases and Increases in Real Income

An index adjustment of the calculation values is necessary to prevent their value being reduced in relation to the price level in the economy as a whole. Of the calculation values taken up in this report the following are adjusted in accordance with CPI: time values for private travel, the "private" portion of business travel values, accident values, noise values and values for air pollution. In the case of carbon dioxide, no adjustment has been made pending the ongoing review of the environmental goals. The portion of the business travel values that reflects the company's profit is adjusted in accordance with changed wage costs. Goods time values are adjusted in accordance with increases in the market values for the respective commodity group. The costs of passenger and goods transport are adjusted in accordance with increases in the respective type of cost.

The calculation values are also adjusted upwards in accordance with increased real income (since the individual's willingness to pay depends on income). No difference is made, however, between different income groups or regions. However, values are adjusted upwards in accordance with increases that have already taken place in *average* income – more exactly in accordance with increase in real GDP per capita from the year in which the relevant valuation study was made to the year whose price level the calculation value is to be expressed in. The calculation values that have been adjusted upwards in this report in accordance with this principle are: time values for private travel, the private portion of the time values for business travel, accident values, noise values and values for the health effects of air pollution.

Goods time values are based on market prices and are not to be adjusted in this way. The same applies to the business portion of business travel values and the costs for passenger and goods transport. No adjustments are made for carbon dioxide. Note also that *no* upward adjustment of the calculation values has been made with regards to increases in income during the calculation period.

Discount Rate, Lifetimes and Tax Factors are not Adjusted

In the previous ASEK review, it was recommended that the discount rate should not include any compensation for uncertainties in the calculation and project risks. This recommendation is retained. Neither are there sufficiently strong reasons for changing the level of the discount rate. The recommended discount rate is thus 4 per cent. In addition, it is recommended that a standard corresponding to 7 per cent be applied as a business financing/interest expense, to be used in business profitability calculations (the discount rate of 4 per cent is used as before for commercial items in the cost benefit calculation).

Lifetimes are unchanged with the exception that the National Road Administration never applies lifetimes exceeding 40 years for bypasses and that all other lifetimes are reported in the national plans together with the net present value calculations. If the lifetimes are longer than the calculation period, the residual value can be taken up in the calculation. The tax factors are also retained at the existing levels, *i.e.* tax factor I is 23 per cent and tax factor II is 30 per cent.

The Valuation of Time in Passenger Transport is not Adjusted although it is a Prioritised Task to obtain New Values in the Near Future

There are a number of reasons why time values for passenger transport should be corrected. Existing empirical (data) for both private and business travel indicate, for instance, that a higher value should be given to delay time (or be valued in the event of it not being valued today) and that congestion time should be valued separately to take into account the additional costs that journeys in congested conditions give rise to. For business time values, there are also issues of principle attached to the choice of valuation approach that have to be clarified and which can lead to changed values. However, the basis for producing concrete material for new values has been considered to be too weak.

In the short term, it is recommended that a new review be made to consider adjustment of the parameter values already included in current valuation approaches and to make supplements for congestion and delays. It is intended that more reliable calculation values should be produced *before* the beginning of the next direction planning. In order for better values to be produced in the long term, it is important that new basic research is done in the area. Until new underlying material is available, it is recommended that the previous calculation values are adjusted according to the CPI index and adjusted upwards in accordance with real GDP per capita.

Small Adjustments are Made in the Valuation of Time and Quality in Goods Transport–Priority is to be given to Produce New Values in the Near Future

In this field discussions have mainly concerned the valuation approach that is to be applied, if, for instance, the capital value method is sufficient to capture all relevant costs or whether there are logistic effects or similar effects which have to be valued separately. Other issues that have been discussed are the present application of a cost of capital of 20 per cent, valuation of secondary time gains in the handling system, valuation of punctuality, valuation of changed delay risk, valuation of changed damage risk and additional demands for cost of capital relating to, for instance, new goods group categorisation. However, no major changes in costs of capital are recommended.

As regards costs of capital for goods time, only a "technical change" is recommended, which entails an adaptation of present calculation values to the new goods groups and the new base and forecast years. It is moreover recommended that the calculated goods time values be multiplied by tax factor I. Regarding the calculation values for changed delay risk, it is recommended that these be adapted to new commodity groups as well including tax factor I. Previous calculation values for delay time are no longer applied. In addition, it is proposed that the ongoing work of reviewing the calculation values be continued with the aim of producing more calculation values for *delays* (time and/or risk) *before* the start of the next direction planning.

The Valuation of Traffic Safety has not been Adjusted–New Basic Research is Needed to Produce more Reliable Values

The material that has been produced in this area in recent years does not provide sufficient support for adjusting accident values. However, current values are generally considered to be very uncertain since the studies the values are based on are still associated with problems at both the principle and practical level. The principal problems are, inter alia, related to the fact that it is very difficult for an individual to understand the meaning of reducing an already very slight risk for something as serious as fatalities, and also knowing how much this changed risk is worth in money. Knowledge is also lacking as to how road users' view of the risks that they expose themselves and others to varies in different situations. Thus, we do not at present have a basis for differentiating accident values in the way that can be needed.

It is therefore recommended that the previous values be retained, apart from an index adjustment according to CPI and an upward adjustment in accordance with increase in real GDP per capita. To be able to produce more reliable values in the future, which better reflect the road users' valuations of accident risks and how these vary in different situations, priority should be given to initiating different types of research and development initiatives in the area.

The Valuation of Noise is not Adjusted–New Research is Needed to Clarify the Existing Correlations for Disturbance

There is a need to develop the existing noise valuations in several respects. Above all, it can be important to evaluate noise in the areas that at present lack valuations, which is the case, for instance, for various work and recreation environments. However, it has not been possible to produce any proposals for new valuations. It is recommended though that the present valuations are applied to evaluate noise disturbance in work and recreation areas in the same way as they are applied to evaluate noise disturbance in residential areas. It is also recommended that an index adjustment be made in accordance with CPI, and an adjustment in accordance with increases in real GDP per capita.

All valuations applied today are based wholly or partly on the valuation studies made for road traffic in residential environments. The disruption correlations can, however, vary between residential, educational, work and recreation environments, and between modes of transport. Producing better knowledge of how these correlations vary is a prerequisite to be able to produce valuations that better reflect how people experience noise in different situations. Other studies can also be important, for instance, to investigate the health effects that noise disturbances can lead to which the individual is unaware of.

No Adjustments are to be made in the Valuation of Air Pollution–However, a New Valuation Approach is to be Applied in the Next ASE Kreview

With respect to particle and NOx valuations, no adjustments are made pending clarification of some remaining questions. A study should therefore be initiated to obtain a relevant and accurate valuation of NOx. Since there is a risk for double counting with regards to the health effects of particles and NOx, consideration should be given at the same time as to whether an adjustment of the NOx valuation should lead to an adjustment of the valuation of the effects on health of particles. Until new material is available, it is only recommended that an adjustment be made in accordance with CPI and in accordance with an increase in real GDP per capita.

Regarding future reviews, it is moreover recommended that the valuation approach applied today should be replaced by the so-called ExternE-model. The reason is that the model will be increasingly normative in the international work of estimating marginal costs for the environmental effects of traffic and will also be used in Sweden to an increasing extent. It is also appropriate to use the same valuation for all modes of transport and for different documentation for transport policy decisions. However, it is important that a review is made at quality assurance of ExternE values based on Swedish conditions before applying ExternE-based values.

Carbon dioxide valuation will be reconsidered when the ongoing review of the subsidiary transport policy objective for carbon dioxide has been carried out

The present carbon dioxide value is based on the current carbon dioxide subsidiary objective for the transport sector. The value will therefore be reconsidered only when the current review of this subsidiary objective has been carried out. However, this does not imply that the new value will be based on the new subsidiary objective, or remain unchanged if the review does not lead to any new subsidiary objective.

The starting point proposed by SIKA to apply to establish a new value is that it should be linked to the *actual Swedish* climate policy ambitions. An alternative is therefore for the valuation to be based on an estimated cost to achieve the currently established climate policy goal. Another alternative is to base the valuation on a *revised* calculation of the costs to achieve the carbon dioxide subsidiary objective for the transport sector, which is accordingly to take place in conjunction with the present review being carried out. If the latter alternative is adopted, the carbon dioxide value can then be changed even if the subsidiary objective is retained. Basing the valuation on Swedish ambitions also means that the establishment of such a value in ASEK should not be an obstacle for Swedish authorities to use values that are more relevant for the context in cases where international consideration *must* be taken.

New Costs have been Produced for Passenger Transport

It has not been motivated to carry out a new survey to update costs in bus and coach traffic. These have only been adjusted according to index. With regards to car traffic, certain adjustments have been made. New values have been produced for the new car price, fuel prices have been adjusted upwards taking into account data from Statistics Sweden and the Swedish Petroleum Institute, tyre costs have been adjusted upwards taking into account the total price information from tyre chains in Sweden. A wage cost is also proposed.

With respect to air transport, new costs have been produced for the items Fixed cost distance, Marginal cost time, Marginal cost distance, Marginal cost time and Capacity use. In the case of train transport, it has not been possible to adopt and process the survey on which the new values were to be based, which has meant that only an index adjustment has been made in this area. New values will be produced when the new survey has been adopted. Costs for shipping have not been considered in the review.

Several Changes have been Made in the Costs for Goods Traffic

The recommendations for new calculation parameters for costs in goods traffic contain a number of changes in comparison with previous values. A new mode of transport has been produced to reflect cars in commercial traffic, and the method of calculation for the transport costs of shipping has been revised. Another new development is that costs for air transport have been produced and some calculation parameters that were previously used but not presented have been highlighted. In addition to this, a new mode of transport has been introduced in Samgods/Samkalk, lorries without a trailer. The number of commodity groups and the number of vehicles with a trailer in Samgods/Samkalk has moreover been increased. The new calculation parameters have also been increased to 2001's price level.

No Supplements in the Calculations for Taking the Regional Economic Effects of Infrastructural Measures into Consideration

New infrastructure can have important effects on regional development. The larger part of these benefits has been captured, however, by the analytical tools applied in the transport sector. At the same time, it is well known that the existing analytical tools do not capture all the effects on the regional economy. However, the assessment is made that the *additional* effects that arise on top of those that are captured in traditional calculations are small for the great majority of measures. There is still a lack of good tools to quantity the effects.

It is therefore recommended that additional benefits should not normally be added to the calculations. For measures where the additional effects can nevertheless be marked, a description of such expected effects is, however, an important part of the cost-benefit assessment. For measures where distribution policy aspects are significant, it is important to report distribution effects even though they are not to be taken into account in the traditional calculation.

It has not been possible to establish calculation values for the effect of the infrastructure on the natural and cultural environment

There is a lack of material at present to produce preference-based values for encroachment effects of a kind that could be used in cost-benefit analyses. Since encroachment effects are very heterogeneous, almost specific to situations, it is also an open question whether the values based on highly simplified assumptions of the homogeneity of various encroachment effects would really provide significant information in the underlying material for decision-making.

However, there is a point in beginning the development of a structure for sorting estimated encroachment values to capture the ranges of sizes for different kinds of effects. The cost-benefit calculations that the transport agencies carry out could then contain a calculation for the specific project design that it is ultimately decided to recommend. It should be possible to achieve this without considerable additional costs. The element of cost-benefit analysis which does not assume an economic valuation of the encroachment could be developed in this way.

It would also be of value if the transport agencies were able to provide a systematic account of the additional (or reduced) costs to society associated with different project designs with typically different degrees of encroachment. In this way, a knowledge base can be built up which would eventually show how encroachment of different kinds and extent was valued *de facto*. An analysis of material of this kind could also be used to determine the interval for valuation of different kinds of encroachment with the aid of which it should be possible to obtain an at least rough valuation of the residual encroachment.

Development inputs are Needed to Improve Application of Cost Benefit Methods in the Area of Operation and Maintenance

Operation and maintenance activity account for approximately half of the National Rail Administration and the National Road Administration's appropriations. The activity is difficult and complex, as well-developed in technical terms as investment activities but associated with considerable deficiencies with regards to effect correlation and valuations as well as modelling tools to enable different kinds of analyses.

At present, a major deficiency is the absence of documented goal standards based on cost benefit assessments. Such goal standards should be developed. For these to be broadly used, it is also important to use user-friendly tools which in turn makes demands on users' competence.

At the same time as producing good methods and tools, continuity and competence must be increased over a very large and complicated area. In the present situation, the lack of quality-assured effect correlations also represents a big deficit. Investment is therefore required in production, further development and quality assurance of knowledge in many areas. In addition, additional investigations are required to produce calculation values that reflect the traffic users' valuations of improved standards on roads and rail.

Sensitivity Analyses as a Method for Handling Uncertainty and Risk in Strategic Planning

An important issue is how the risks associated with investments should be handled in the light of the uncertainty that exists on surrounding world conditions and on the outcome of costs, passenger numbers and goods transport volumes. The recommendation that is made is that these uncertainties should in the first place be dealt with at the strategic level and that this is to be done by sensitivity analyses where net social benefit of packages of measures are examined.

Uncertainty about the outcome of the investment costs is analysed by comparing the net social benefit obtained when using the calculated cost with the net social benefit obtained when using the calculated cost plus a measure of expected (in a statistical meaning) discrepancy based on previous historical discrepancies between calculated and actual cost. In a similar way, the uncertainty of the outcome of passenger numbers and goods transport volumes is analysed by comparing the net social benefits obtained in different main scenarios. More important calculation values such as carbon dioxide value, petrol price, time values and risk values can be made the subject of uncertainty analyses in the strategic planning phase. Certain sensitivity analyses should also be carried out or individual items but to a considerably smaller extent.

There are Correlations between Measures that are Important to take into Consideration

Sometimes the net social benefit of a particular measure can be highly dependent on *e.g.* the other measures that have been carried out at the same time. To clarify the importance of taking these correlations in cost-benefit calculations into consideration, we have analysed three areas in this report where there is a large mutual impact on the effects (or net benefit) of measures. These are road safety measures, roads in big cities and the correlations between investments.

The discussions in the report lead to some concrete recommendations. One is that it should be shown in a strategic analytical phase how net social benefits of investments in new roads, reconstruction of existing roads, and other targeted traffic safety measures are affected by being calculated on the basis of current speeds or with "optimal" speeds, *i.e.* speeds that minimise the total use of society's economic resources. Another recommendation is that net social benefits should always be calculated for major road projects with and without charges based on social marginal costs (The National Road Administration have made a reservation against the proposal that charges be based on marginal costs – it is more relevant, it is said, to base the calculations on *planned* road charges). A third recommendation is that the correlation between different investments should be analysed for a selection of cases, for both the correlation between different distances on the same route and between different routes.

No Standardised Benefit Amounts Should be Added to Costs as a Method for Handling Utilities and Costs that are Difficult to Value

At present, there is no general recommendation on how utilities and costs that are difficult to value are to be dealt with. The National Rail Administration has in its calculation manual adopted a very restrictive approach, while the National Road Administration proposed an approach in *Effektsamband* 2000 which entails that a standardised benefit corresponding to the additional cost plus tax factors can be added to the calculations for measures that are associated with utilities that are hard to value.

In this report, SIKA has argued that such utility standards should *not* be added to the calculations. SIKAs position is that all utilities that are added to the calculations should in principle be derived from studies that aim at clarifying the utility of the measures in terms of willingness to pay.

It is Important to Design the Investment Calculations so that thay can be Followed Up

A study made by SIKA shows that it is difficult for railway investments to make a correct comparison between the traffic that is included in the net social benefit calculation and the traffic that actually takes place some time after the investment has been completed. An important reason for this is that all investments in a plan are calculated as if they were started on the first day of the plan period. It is therefore important to change the design of the calculations so that they can be followed up.

SIKA therefore proposes the following procedure: For major projects (larger than SEK 1 billion) that are constructed in stages, an assumption should be made about the order in which the stages will be built. Thereafter an attempt should be made to describe a conceivable course of events for the adaptation of the volume of traffic to the expansion in capacity. Given a description of the development of the transport offered, a simplified forecast of how passenger volumes will develop can be made. This forecast can be made as an interpolation between present travel volume and the travel volume in the forecast year with the aid of elasticity assessments. A calculation should then be made for this realistic development of the traffic volumes.

Clearer Demands Should be Made on Documentation of Forecasts

A major infrastructure project is valued during its planning, as a part of different packages, typically with a number of different forecasts. It has proven difficult retrospectively to recreate these various forecasts. This is related to the forecast models being successively further developed and the documentation of input data and forecast prerequisites not being sufficiently extensive and systematic. The same problem applies for the calculations carried out at different stages in the process. It would therefore be apposite if SIKA and the traffic agencies together worked out guidelines for how the prerequisites and forecasts as well as calculations and their prerequisites could be documented.

The Cost Benefit Analysis can be Modified to Comply Better with the Current Decision-Making Situation

Traditional cost-benefit calculation methods based on valuations derived from the individual preferences of citizens constitute a scientifically well-founded, wellknown and tested method for producing a basis for decision-making. When the cost-benefit method is used in the transport sector, a traditional arrangement of the cost-benefit analysis should therefore be applied. In certain well-defined cases, however, it may be necessary to expand or modify the cost-benefit analysis to correspond better to the current decision-making situation.

In a situation where political decisions have been made on the balances to be struck that constitute a starting point for the analysis, it may accordingly be more relevant to design the cost-benefit analyses so that they indicate *which social costs can be associated with achieving particular goals* rather than those that are expected to be effective in a cost-benefit sense. In a planning or decision-making situation based on already given and highly detailed goals, it should be possible to adjust the monetary values on which the cost-benefit calculations are based so that they correspond to the stated goals. A prerequisite is that the politically given restrictions for the cost-benefit analyses are very clearly stated in the planning directive or equivalent.

In a traditional cost-benefit calculation, valuations derived from political decisions can never be regarded as a fully satisfactory replacement for valuations that are derived from citizens' individual preferences. When the prerequisites are lacking to obtain valuations of the latter kind, it should be possible in certain cases to use the values derived from binding political decisions of the latter kind to make the calculations more complete.

Chapter 10

Economics of Biodiversity

Biodiversity is the variability of all living organisms – including animal and plant species – of the genes of all these organisms, and of the terrestrial, aquatic and marine ecosystems of which they are part.

Biodiversity makes up the structure of the ecosystems and habitats that support essential living resources, including wildlife, fisheries and forests. It helps provide for basic human needs such as food, shelter, and medicine. It composes ecosystems that maintain oxygen in the air, enrich the soil, purify the water, protect against flood and storm damage and regulate climate. Biodiversity also has recreational, cultural, spiritual and aesthetic values.

Society's growing consumption of resources and increasing populations have led to a rapid loss of biodiversity, eroding the capacity of earth's natural systems to provide essential goods and services on which human communities depend. Human activities have raised the rate of extinction to 1,000 times its usual rate. If this continues, Earth will experience the sixth great wave of extinctions in billions of years of history. Already, an estimated two of every three bird species are in decline worldwide, one in every eight plant species is endangered or threatened, and one-quarter of mammals, one-quarter of amphibians and one-fifth of reptiles are endangered or vulnerable.

Also in crisis are forests and fisheries, which are essential biological resources and integral parts of the earth's living ecosystems. The World Resources Institute estimates that only one-fifth of the earth's original forest cover survives unfragmented, yet deforestation continues, with 180 million hectares in developing countries deforested between 1980 and 1995. Forests are home to 50-90 per cent of terrestrial species, provide ecosystem services such as carbon storage and flood prevention, and are critical resources for many linguistically and culturally diverse societies and millions of indigenous people.

Over fishing, destructive fishing techniques and other human activities have also severely jeopardized the health of many of the world's fish stocks along with associated marine species and ecosystems. The Food and Agriculture Organization of the UN estimates that nearly two-thirds of ocean fisheries are exploited at our beyond capacity. Over one billion people, mostly in developing countries, depend on fish as their primary source of animal protein.

Key Terms

☆ An ecosystem is a dynamic complex of plant, animal and micro-organism communities and their non-living environment interacting as a functional unit. Examples of ecosystems include deserts, coral reefs, wetlands, rainforests, boreal forests, grasslands, urban parks and cultivated farmlands. Ecosystems can be relatively undisturbed by people, such as virgin rainforests, or can be modified by human activity.

☆ Ecosystem services are the benefits that people obtain from ecosystems. Examples include food, freshwater, timber, climate regulation, protection from natural hazards, erosion control, pharmaceutical ingredients and recreation.

☆ Biodiversity is the quantity and variability among living organisms within species (genetic diversity), between species and between ecosystems. Biodiversity is not itself an ecosystem service but underpins the supply of services. The value placed on biodiversity for its own sake is captured under the cultural ecosystem service called "ethical values".

Biofuels Generate much Debate

Biofuels generate much debate Bioenergy can play an important role in combating climate change, specifically if biomass is used for heat and electricity generation. However, biofuels also are another source of competition for scarce land, and the scale of potential land conversion for agro-fuels is extraordinary. The International Monetary Fund reports that "although biofuels still account for only 1.5 per cent of the global liquid fuels supply, they accounted for almost half of the increase in consumption of major food crops in 2006-07, mostly because of corn-based ethanol produced in the US". Reports indicate that this pattern could be replicated elsewhere in the world.

Coral Reefs

Coral reefs are the most biodiversity-rich ecosystems (in species per unit area) in the world, more diverse even than tropical forests. Their health and resilience are in decline because of over fishing, pollution, disease and climate change.

Caribbean coral reefs have been reduced by 80 per cent in three decades. As a direct result, revenues from dive tourism (close to 20 per cent of total tourism revenue) have declined and are predicted to lose up to US$ 300 million per year. That is more than twice as much as losses in the heavily impacted fisheries sector (UNEP February 2008).

The underlying explanation for this situation is that in 1983, following several centuries of over fishing of herbivores, there was a sudden switch from coral to algal domination of Jamaican reef systems. This left the control of algal cover almost entirely to a single species of sea urchin, whose populations collapsed when exposed to a species-specific pathogen. When the sea urchin population collapsed, the reefs shifted (apparently irreversibly) to a new state with little capacity to support fisheries. This is an excellent example of the insurance value in biologically diverse ecosystems. The reduction in herbivore diversity had no immediate effect until the sea urchin population plummeted, illustrating how vulnerable the system had become due to its dependence on a single species.

Our Health is at Stake

People have known the medicinal value of certain plants for thousands of years and biodiversity has helped our understanding of the human body. So ecosystems provide huge health benefits, and thus economic benefits. The corollary is that losing biodiversity incurs potentially huge costs, and our knowledge of these is growing (Conseil Scientifique du Patrimoine Naturel et de la Biodiversité – in press).

There are significant direct links between biodiversity and modern healthcare (Newman and Cragg, 2007):

☆ Approximately half of synthetic drugs have a natural origin, including 10 of the 25 highest selling drugs in the United States of America.

☆ Of all the anti-cancer drugs available, 42 per cent are natural and 34 per cent semi-natural.

☆ In China, over 5,000 of the 30,000 recorded higher plant species are used for therapeutic purposes.

☆ Three quarters of the world's population depend on natural traditional remedies.

☆ The turnover for drugs derived from genetic resources was between US$ 75 billion and US$ 150 billion in the United States of America in 1997.

☆ The gingko tree led to the discovery of substances which are highly effective against cardiovascular diseases, accounting for a turnover of US$ 360 million per year.

Despite the enormous health benefits, plants are disappearing fast and will continue to do so unless urgent action is taken. The *2007 IUCN Red List of Threatened Species* identified a significant increase in species under threat during this decade. It estimates that 70 per cent of the world's plants are in jeopardy (IUCN, 2008).

A recent global study reveals that hundreds of medicinal plant species, whose naturally occurring chemicals make up the basis of over 50 per cent of all prescription drugs, are threatened with extinction. This prompted experts to call for action to "secure the future of global healthcare". (Hawkins, 2008).

The biodiversity-healthcare relationship also has a strong distributional equity dimension. There is often a mismatch between the regions where benefits are produced, where their value is enjoyed, and where the opportunity costs for their conservation are borne. So the plant species that are the sources of many new drugs are likely to be found in poorer tropical regions of the world. The people that benefit are more likely to be found in rich countries where the resulting drugs are more readily available and affordable. People in these countries therefore have a great incentive to conserve natural habitats in biodiversity-rich parts of the world. However, such conservation has costs for local people in these parts, in particular the opportunity costs such as the loss in potential agriculture returns of not converting such habitats. Transferring some of the rich world benefits back to local people could be one approach to improving incentives to conserve those natural habitats and species locally that clearly have wider benefits globally.

It is clear that if we undermine the natural functions that hold this planet together, we may be creating conditions that will make life increasingly difficult for generations to come – and impossible for those already on the margins of survival.

Growth and Development

Population growth, increasing wealth and changing consumption patterns underlie many of the trends we have described. Unsustainable resource use has been evident in the developed world for many years. The ecological footprints of Europe, the United States of America and Japan are much higher than those of developing countries. And the emerging economies are catching up. India and China both have ecological footprints twice the size of their "biocapacities" (Goldman Sachs 2007) – the extent to which their ecosystems can generate a sustainable supply of renewable resources. Brazil, on the other hand, has one of the world's highest "biocapacities", nearly five times as large as its ecological footprint, yet this is declining as a result of deforestation (Goldman Sachs, 2007).

Under current practices, meeting the food needs of growing and increasingly affluent populations will further threaten biodiversity and ecosystem services. Based on population projections alone, 50 per cent more food than is currently produced will be required to feed the global population by 2050 (United Nations Department of Economic and Social Affairs/Population Division, 2008). Irrigated crop production will need to increase by 80 per cent by 2030 to match demand.

Already, 35 per cent of the Earth's surface has been converted for agriculture, limiting scope for the future productivity of natural systems. The livestock sector already represents the world's single largest human use of land. Grazing land covers 26 per cent of the Earth's surface, while animal feed crops account for about a third of arable land. Extending agricultural production will have consequences for biodiversity and ecosystem services as more land is converted for food production. The expanding livestock sector will be in direct competition with humans for land, water and other natural resources. Livestock production is the largest sectoral source of water pollutants. It is also a major factor in rising deforestation: 70 per cent of land in the Amazon that was previously forested is now used as pasture, and livestock feed crops cover a large part of the remainder (FAO, 2006).

Climate Change and Biodiversity

Climate change is linked to many of the issues we have presented in this chapter. The El Nino-La Nina cycle in the Pacific Ocean is one prominent example of the vulnerability of biodiversity to climate. A small rise in the sea surface temperature in 1976 and 1998 led to a series of worldwide phenomena, which resulted in 1998 being characterized as "the year the world caught fire". Permanent damage includes (US Department of Commerce, 2008):

☆ Burned forests that will not recover within any meaningful human timescale;

☆ A rise in the temperature of surface waters of the central western Pacific Ocean from an average of 19°C to 25°C;

☆ Shifts toward heat-tolerant species living inside corals;

☆ A northward shift in the jet stream.

These types of complex phenomena show us how vulnerable we are to tipping points beyond those linked directly to increasing temperatures and carbon dioxide levels.

Biodiversity losses can also contribute to climate change in many complex ways. There are many examples of how overharvesting or changed land-use patterns have triggered social and economic changes leading to greater reliance on carbon.

Draining peat lands results in carbon losses. But predicted changes to climate could cause accelerated rates of carbon release from the soil, contributing in turn to higher greenhouse gas concentrations in the atmosphere (Bellamy *et al.*, 2005). Under the same climatic conditions, grassland and forests tend to have higher stocks of organic carbon than arable land and are seen as net sinks for carbon. Yet deforestation and intensification of cropland areas are rampant.

To take account of these complexities we will need more than energy-based econometric models. We will need to respond to knowledge about how to adapt and how vulnerabilities might arise from global ecological processes. This will require a much deeper dialogue than we have seen so far between economists, climate scientists and ecologists.

Impacts on the Poor

A striking aspect of the consequences of biodiversity loss is their disproportionate but unrecognized impact on the poor. For instance, if climate change resulted in a drought that halved the income of the poorest of the 28 million Ethiopians, this would barely register on the global balance sheet – world GDP would fall by less than 0.003 per cent. The distributional challenge is particularly difficult because those who have largely caused the problems – the rich countries – are not going to suffer the most, at least not in the short term.

The evidence is clear. The consequences of biodiversity loss and ecosystem service degradation – from water to food to fish – are not being shared equitably across the world. The areas of richest biodiversity and ecosystem services are in developing countries where they are relied upon by billions of people to meet their basic needs. Yet subsistence farmers, fishermen, the rural poor and traditional societies face the most serious risks from degradation. This imbalance is likely to grow. Estimates of the global environmental costs in six major categories, from climate change to over fishing, show that the costs arise overwhelmingly in high- and middle-income countries and are borne by low-income countries (Srinivasan *et al.*, 2007).

Gender, Poverty and Biodiversity in India

The impact of the loss of biodiversity, often not very visible, has serious implications for poverty reduction and well-being for women as it severely affects the role of women as forest gatherers. Studies in the tribal regions of Orissa and Chattisgarh, states in India which were once heavily forested, have recorded how deforestation has resulted in loss of livelihoods, in women having to walk four times the distance to collect forest produce and in their inability to access medicinal herbs which have been depleted. This loss reduces income, increases drudgery and affects physical health. There is also evidence to show that the relative status of women within the family is higher in well-forested villages, where their contribution to the household income is greater than in villages that lack natural resources.

Business-as-Usual is not an Option

If no major new policy measures are put in place, past trends of biodiversity and ecosystem service loss will continue.

The changing use of land and changing services Humans have been causing biodiversity loss for centuries (see maps below). By the year 2000, only about 73 per cent of the original global natural biodiversity was left. The strongest declines have occurred in the temperate and tropical grasslands and forests, where human civilizations first developed (McNeill and McNeill 2003).

A further 11 per cent of land biodiversity is expected to be lost by 2050, but this figure is an average including desert, tundra and polar regions. In some biomes and regions, projected losses are about 20 per cent. Natural areas will continue to be converted to agricultural land, with the ongoing expansion of infrastructure and increasing effects of climate change being additional major contributors to biodiversity loss. For the world as a whole, the loss of natural areas over the period 2000 to 2050 is projected to be 7.5 million square kilometres or around 750 million hectares, *i.e.* the size of Australia. These natural ecosystems are expected to undergo human-dominated land-use change in the next few decades. Biodiversity loss in the Cost of Policy Inaction (COPI) study is measured by the MSA (mean species abundance) indicator, a reliable measure of biodiversity that has been recognized by the Convention on Biological Diversity.

The impact on livelihoods is local and therefore not necessarily reflected in aggregate global numbers. Maps can give a clearer picture and the figures below show the changes in biodiversity based on mean species abundance between 1970, 2000, 2010 and 2050. Major impacts are expected in Africa, India, China and Europe (Braat, ten Brink *et al.*, 2008).

Some cases losses will accelerate. In others the ecosystem will be degraded to such an extent that it will not be possible to repair or recover it. These are some of the likely results of inaction:

☆ Natural areas will continue to be converted to agricultural land, and will be affected by the expansion of infrastructure and by climate change. By 2050, 7.5 million square kilometres are expected to be lost, or 11 per cent of 2000 levels (see next section) (Braat, ten Brink *et al.*, 2008).

☆ Land currently under extensive (low-impact) forms of agriculture, which often provides important biodiversity benefits, will be increasingly converted to intensive agricultural use, with further biodiversity losses and with damage to the environment. Almost 40 per cent of land currently under extensive agriculture is expected to be lost by 2050 (Braat, ten Brink *et al.*, 2008).

☆ 60 per cent of coral reefs could be lost by 2030 through fishing damage, pollution, disease, invasive alien species and coral bleaching, which is becoming more common with climate change. This risks losing vital breeding grounds as well as valuable sources of revenue to nations (Hughes *et al.*, 2003).

☆ Valuable mangrove areas are likely to be converted to use for private gain, often to the detriment of local populations. Important breeding grounds will be lost, as will buffers that protect against storms and tsunamis.

☆ If current levels of fishing continue, there is the risk of collapse of a series of fisheries. The global collapse of most world fisheries is possible by the second half of the century unless there is an effective policy response – and enforcement (Worm *et al.*, 2006).

☆ As global trade and mobility increase, so do the risks from invasive alien species for food and timber production, infrastructure and health.

Business-as-usual is not an option if we wish to avoid these consequences and to safeguard our natural capital and the well-being of future generations. The cost of insufficient policy action is too great.

Some solutions are already visible, however, and economics could play an important part. Although forests are at risk of conversion to agriculture, grazing lands and biofuel production, they can play a valuable role as carbon sinks and biodiversity vaults, and this capacity could be recognized by a higher market value.

Managing humanity's desire for food, energy, water, lifesaving drugs and raw materials, while minimizing adverse impacts on biodiversity and ecosystem services, is today's leading challenge for society. Maintaining an appropriate balance between competing demands means understanding economic resource flows and tracking the biological capacity needed to sustain these flows and absorb the resulting waste.

The Effect of Subsidies on Fisheries

Subsidies are considered to be one of the most significant drivers of over fishing and thus indirect drivers of degradation and depletion in marine biodiversity.

☆ Subsidies fund fisheries expansion. Globally, the provision of subsidies to the fisheries industry has been estimated at up to US$ 20-50 billion annually, the latter roughly equivalent to the landed value of the catch.

☆ Over half the subsidies in the North Atlantic have negative effects through fleet development. This includes decommissioning subsidies, which have been shown usually to have the effect of modernizing fleets, thereby bringing about an increase in their catching powers.

☆ While fishing vessel populations stabilized in the late 1990s, cheap fuel subsidies keep fleets operating even when fish are scarce.

☆ The Common Fisheries Policy of the European Community, for example, allows for vessels to be decommissioned to reduce effort in some countries while simultaneously subsidizing others to increase their fishing capacity.

Recognizing Risks and Uncertainty

The treatment of climate change by the *Stern Review* surfaced an issue which had been widely recognized but not tackled squarely: how to assess a roll of the dice, when one of the outcomes is the end of civilization as we know it?

This dilemma also applies to assessing the risks of ecosystem collapse. The difficulty was highlighted when one academic study (Costanza *et al.* 1997) estimated the economic value of ecosystem services at US$ 33 trillion (compared to US$ 18 trillion for global GDP). This result was criticized on the one hand for being far too high, but on the other hand for being "a significant underestimate of infinity" (Toman 1998).

Expressed in the language of finance, the global economy is "short an option" on climate change and on biodiversity and needs to pay a premium to buy protection. The *Stern Review*'s most quoted result, that a 1 per cent per annum cost would be needed to protect the world economy from a loss of up to 20 per cent of global consumption, is an example of such an "option premium".

In the case of biodiversity and ecosystem losses, the size of such premiums will depend on several aspects of the ecosystem in question: its current state, the threshold state at which it fails to deliver ecosystem services, its targeted conservation state, and our best estimate of uncertainties. This is an exceedingly complex exercise as there are no market values for any of these measures.

Discount Rates and Ethics

We are addressing issues here (such as species extinction) where there is no universal agreement on the appropriate ethics. But the ethical nature of the issue is widely recognized. A group of ethics experts (IUCN Ethics Specialist Group, 2007) recently framed the issue like this:

"If human behaviour is the root cause of the biodiversity extinction crisis, it follows that ethics – the inquiry into what people and societies consider to be the right thing to do in a given situation – must be part of the solution. However, ethics is rarely accepted as an essential ingredient and is usually dismissed as being too theoretical a matter to help with the urgent and practical problems confronting conservationists."

Table 1: Valuing "biodiversity option"

Measures of		Financial Option	Biodiversity Option
(a)	Current value	Spot price	All variables–Current state
(b)	Level of protection	Strike price	All variables–Future state
(c)	Life of protection	Expiration	Conservation horizon
(d)	Uncertainty	Implied volatility	Modelled uncertainty
(e)	Discounting	Interest rate	Social discount rate

This analogy with a financial option illustrates how complex it would be to price a "biodiversity option". All five input variables (a) to (e) for a financial option have market values, as agaist NONE of these for biodiversity.

Economists discount any future benefit when comparing it to a current benefit. At one level, this is just a mathematical expression of the common-sense view that a benefit today is worth more than the same benefit in the future. But ethical considerations arise, for example when we consider giving up current income for the benefit of future generations, or the opposite: gaining benefits now at the expense of future generations.

Financial discount rates consider only the time value of money, or the price for its scarcity, and relate the present value of a future cash flow to its nominal or future value. Simple discount rates for goods and services consider just time preference, or the preference for a benefit today versus later. Social discount rates are more complex, and engage ethical aspects of a difficult choice: consumption now versus later, for society rather than for an individual. The preferences built into this choice cover the relative value of goods or services in the future when their benefit may be lower, or higher, than now, and that benefit might flow to a different person or to a future generation.

Discounting and Intergenerational Equity

The *Stern Review* has highlighted the crucial importance of the choice of discount rates in long-term decisions that range beyond conventional economic calculations. The discount rate has even been described as the "biggest uncertainty of all in the economics of climate change" (Weitzman, 2007).

Discounting and the Optimist's Paradox

There are two main reasons for discounting. The first is called "pure time preference" by economists. It refers to the inclination of individuals to prefer 100 units of purchasing power today to 101, or 105, or even 110 next year, not because of price inflation (which is excluded from the reasoning) but because of the risk of becoming ill or dying and not being able to enjoy next year's income. Whatever the reason for this attitude, it should not apply to a nation or to human society with a time horizon in the thousands or hundreds of thousands of years. Economists have often criticized "pure time preference". The most famous critique against it was perhaps that of the Cambridge economist Frank Ramsey, in 1928.

In the context of growth theory, economists agree with the discounting of the future for other reasons. They might agree with Ramsey, that to discount later enjoyments in comparison with earlier ones is "a practice which is ethically indefensible and arises merely from the weakness of the imagination". But discount they will, as Ramsey himself did, because they assume that today's investments and technical change will produce economic growth. Our descendants will be richer than we are. They will have three, four or even more cars per family. Therefore, the marginal utility, or incremental satisfaction they will get from the third, fourth or fifth car, will be lower and lower. Discounting at the rate at which marginal utility decreases could be ethically justified.

Growth is then the reason for undervaluing future consumption and future enjoyments. Is it also a reason to undervalue future needs for environmental goods and services? It is not, particularly if we think of irreversible events. Economic growth might produce virtual Jurassic Theme Parks for children and adults; it will never resurrect the tiger if and when it goes.

Growth theory is economic theory. It does not take out from the accounts the loss of nature, nor does it exclude from the accounts the defensive expenditures by which we try to compensate for nature's loss (building dykes against sea-level rise induced by climate change, or selling bottled water in polluted areas).

If we try to add up the genuine increase of the economy because of positive technical changes and investments (which nobody would deny), and the loss of environmental services caused by economic growth, the balance would be doubtful. In fact, we step on the issue of incommensurability of values.

Discounting gives rise to "the optimist's paradox". Modern economists favour discounting not because of "pure time preference" but because of the decreasing marginal utility of consumption as growth takes place. The assumption of growth (measured by GDP) justifies our using more resources and polluting more now than we would otherwise do. Therefore our descendants, who by assumption are supposed to be better off than ourselves, perhaps will be paradoxically worse off from the environmental point of view than we are.

This is because the events being considered will happen over periods of 50 years or more, and the effect of choosing different discount rates over such long periods is significant, as Table 2 shows. The effects of only small differences in the discount rate, applied to a cash flow of US$ 1 million in 50 years' time, are dramatic. A zero discount rate means the cost or benefit is worth the same now as it would be in 50 years, but small increases in the rate result in substantial reductions in the present value of the future cash flow. An annual discount rate of 0.1 per cent produces a present value of 95 per cent of the forward cash flow (US$ 951,253). Discounted at 4 per cent, the result is only 14 per cent of the future cash flow – just US$ 140,713.

Table 2: Discount rates and outcomes.

50-year Forward Cash Flow	Annual Dicount Rate Per cent	Present Vale of Future Cash Flow
1,000,000	4	140,713
1,000,000	2	371,528
1,000,000	1	608,039
1,000,000	0.1	951,253
1,000,000	0	1,000,000

Applying a 4 per cent discount rate over 50 years implies that we value a future biodiversity or ecosystem benefit to our grandchildren at only one-seventh of the current value that we derive from it!

If our ethical approach sees our grandchildren valuing nature similarly to our generation, and deserving as much as we do, the discount rate for valuing such benefits over such a time period should be zero. Unlike man-made goods and services which are growing in quantity (hence the argument to discount future units of the same utility), the services of nature are not in fact likely to be produced in larger quantities

in future. Perhaps the discount rate for biodiversity and ecosystem benefits should even be negative, on the basis that future generations will be poorer in environmental terms than those living today, as Paul Ehrlich (2008) has suggested. That raises important questions about present policies which assume significant positive discount rates (Dasgupta, 2001; 2008).When incomes are expected to grow, goods or services delivered later are relatively less valuable (because they represent a smaller part of the future income). This supports the usual, positive discount factor. The opposite holds true when asset values or incomes are expected to fall – future goods and services will become more valuable than now. In the case of biodiversity it is questionable whether it will be equally, more or less available in future, and therefore even the direction of the discount rate is uncertain.

"GDP of the Poor"

The full economic significance of biodiversity and ecosystems does not figure in GDP statistics, but indirectly its contribution to livelihood and well-being can be estimated and recognized. Conversely, the real costs of depletion or degradation of natural capital (water availability, water quality, forest biomass, soil fertility, topsoil, inclement microclimates, etc.) are felt at the micro-level but are not recorded or brought to the attention of policy makers. If one accounts for the agricultural, animal husbandry and forestry sectors properly, the significant losses of natural capital observed have huge impacts on the productivity and risks in these sectors. Collectively, we call these sectors (*i.e.* agriculture, animal husbandry, informal forestry) the "GDP of the poor" because it is from these sectors that many of the developing world's poor draw their livelihood and employment. Furthermore, we find that the impact of ecosystem degradation and biodiversity loss affects that proportion of GDP most which we term the "GDP of the poor".

The end-use of ecosystem and biodiversity valuations in National Income Accounting, either through satellite accounts (physical and monetary) or in adjusted GDP accounts ("Green Accounts") does not of itself ensure that policymakers read the right signals for significant policy trade-offs. A "beneficiary focus" helps better recognize the human significance of these losses. In exploring an example (GAIS project, Green Indian State Trust 2004-08) for this interim report, we found that the most significant beneficiaries of forest biodiversity and ecosystem services are the poor, and the predominant economic impact of a loss or denial of these inputs is to the income security and well-being of the poor. An "equity" focus accentuated this finding even further, because the poverty of the beneficiaries makes these ecosystem service losses even more acute as a proportion of their livelihood incomes than is the case for the people of India at large. We find that the per-capita "GDP of the poor" for India (using 2002/03 accounts and exchange rates) increases from US$ 60 to US$ 95 after accounting for the value of ecological services, and also that if these services were denied, the cost of replacing lost livelihood, equity adjusted, would be US$ 120 per capita – further evidence of the "vicious cycle" of poverty and environmental degradation.

We shall explore this approach for the developing world more broadly in Phase II. We believe that by using such sectoral measures and forcing a reflection of the equity principle by its "human" significance (given that most of the world's 70 per cent poor are dependent on this sector) we shall focus adequate importance on policy making and contribute to a halt in the loss of biodiversity and drought losses. It could be ethically difficult to justify destroying such a forest watershed in order to release economic value which has utility for the agents of destruction (*e.g.*, profits from minerals and timber, related employment, etc.), whilst on the other hand, the costs of replacing ecosystem benefits forgone may be the same or less in monetary terms, but impossible to bear in human terms as they fall on poor subsistence farming communities. We see such situations as outcomes of bad economic targeting – *economics is mere weaponry, its targets are ethical choices*.

Discounting Biodiversity Losses

We do not suggest that there are always defensible "tradeoffs" for ecosystems and biodiversity, especially if significant ecosystems cease to function altogether as providers of provisioning or regulating services, or if biodiversity suffers significant extinctions. The evaluation of trade-offs using cost-benefit analysis and discounting works best for marginal choices involving small perturbations about a common growth path. However, the reality is that there are trade-offs, explicit or implicit, in any human choice. Even trying to set a boundary where trade-offs should not apply is itself a trade-off!

Trade-offs involve a choice between alternatives, and in the case of biodiversity losses, there are not always comparable alternatives. For development to be considered sustainable, a boundary condition called "weak sustainability" is defined, being a situation in which overall capital – natural, human and physical – is not diminished. But this also suggests that one form of capital can be substituted for another, which is not true: more physical wealth cannot always be a substitute for a healthy environment, nor vice versa. However, it is important for all aspects of the "natural capital" side of a trade-off at least to be appropriately recognized, valued and reflected in cost-benefit analysis, and even this is not yet being done in most trade-off decisions. There is a different boundary condition called "strong sustainability" which requires no net diminution of natural capital: this is more difficult to achieve, although compensatory afforestation schemes are examples of instruments designed to achieve strong sustainability. Finally, any trade-off has to be ethically defensible, and not just economically sound.

With biodiversity, we are not only considering long-term horizons as we are with climate change. Ecosystem degradation is already extensive and observable, and some of its effects are dramatic – such as the loss of freshwater causing international tension. Significant biodiversity losses and extinctions are happening right now, and flagship species such as the Royal Bengal tiger in India are under threat. A higher or lower discount rate can change the quantification of the social cost of imminent losses, but it would not alter the nature of the outcomes – loss of vital ecosystem services and valuable biodiversity.

In one of the accompanying papers of Phase I (IUCN, 2008), approximately 200 valuation studies on forests have been examined.

Many of these included some discounting of annuity flows in order to calculate an aggregate value for natural capital. We found that most studies used social discount rates of 3-5 per cent or higher, and that none were below 3 per cent. Our intention in Phase II is to leverage off this body of work, but to recalculate its results with different discounting assumptions.

Thus in Phase II we will propose a conceptual framework for the economics of biodiversity and ecosystem valuation which includes assessments of the sensitivity of ecosystem values to ethical choices. Our intention is to present a discrete range of discounting choices connected to different ethical standpoints, enabling endusers to make a conscious choice.

The Evaluation Challenge

Economic evaluation can shed light on trade-offs by comparing benefits and costs and taking account of risks, and this can be applied to alternative uses of ecosystems. But there are many difficulties, which we set out in this section, and which we will address in Phase II.

Before economic valuation can be applied it is necessary to assess ecosystem changes in biophysical terms. Most benefits provided by ecosystems are indirect and result from complex ecological processes that often involve long lag times as well as non-linear changes (Figure 1). Pressures may build up gradually until a certain threshold is reached, leading to the collapse of certain functions. A typical example is forest die-back caused by acidification. The impacts of pressures on ecosystems, including the role of individual species, the importance of overall levels of biodiversity, the relationship between the physical and the biological components of the ecosystem, and the consequences with regard to the provision of services, are difficult to predict.

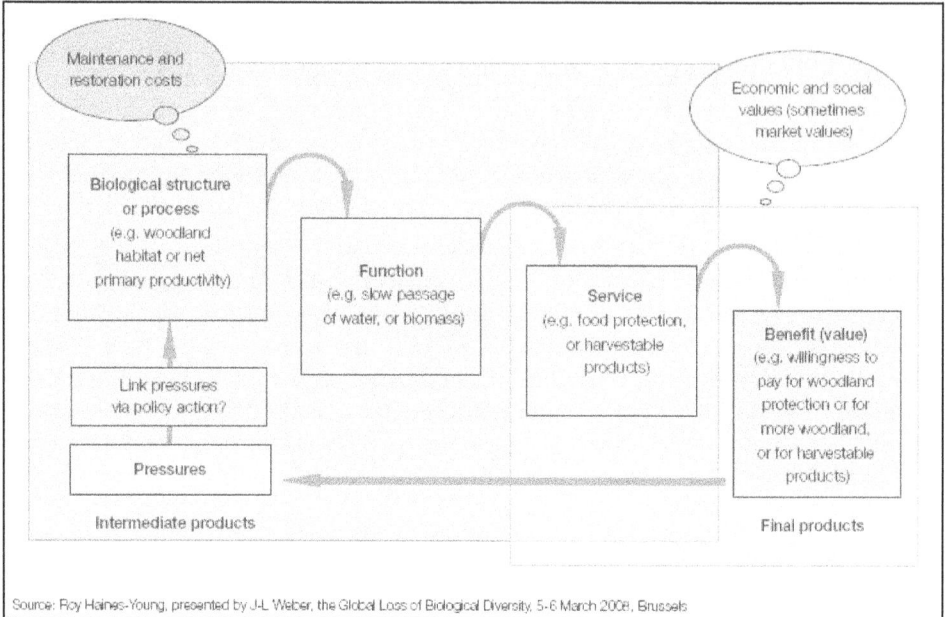

Figure 1: The link between biodiversity and the output of ecosystem services.

Economic valuation builds on the biophysical understanding and aims to measure people's preferences for the benefits from ecosystem processes. These benefits may accrue to different categories of population over different geographical and time scales.

Our ability to assess the benefits provided by ecosystems, or the costs from their loss, is limited by lack of information at several levels. There are probably benefits that we have not yet identified, so we are able to assess, even in qualitative terms, only part of the full range of ecosystem services. We will probably never be able to assess the full range. It will be possible to make a quantitative assessment in biophysical terms only for part of these services – those for which the ecological "production functions" are relatively well understood and for which sufficient data are available. Due to the limitation of our economic tools, a still smaller share of these services can be valued in monetary terms.

Modelling Biodiversity Loss

The GLOBIO model was used to project changes in terrestrial biodiversity to 2050 (OECD, 2008). The main indicators were changes in land use and quality and the mean abundance of the original species of an ecosystem (MSA), for all of the world's biomes. The model provides regional estimates for conversions from natural to managed forest and from extensive to intensive agriculture, and for the resulting decline in natural areas. The largest driver for conversions has historically been demand for agricultural land and timber, although infrastructure development, fragmentation, and climate change are predicted to become increasingly important.

The expected loss of biodiversity by 2050 is about 10-15 per cent (decline in MSA), the most extreme being in savannah and grassland.

It is therefore important not to limit assessments to monetary values, but to include qualitative analysis and physical indicators as well. The "pyramid" diagram in Figure 2 illustrates this important point.

Measurement approaches vary depending on what we measure. For provisioning services (fuel, fibre, food, medicinal plants, etc.), measuring economic values is relatively straightforward, as these services are largely traded on markets. The market prices of commodities such as timber, agricultural crops or fish provide a tangible basis for economic valuation, even though they may be significantly distorted by externalities or government interventions and may require some adjustments when making international comparisons.

For regulating and cultural services, which generally do not have any market price (with exceptions such as carbon sequestration) economic valuation is more difficult. However, a set of techniques has been used for decades to estimate non-market values of environmental goods, based either on some market information that is indirectly related to the service (revealed preference methods) or on simulated markets (stated preference methods). These techniques have been applied convincingly to many components of biodiversity and ecosystem services (an overview of the suitability of these methods to valuate ecosystem services is provided by the Millennium Ecosystem Assessment (2005b). But they remain controversial.

The scenario used was largely developed by the OECD as its baseline (OECD 2008). It is broadly consistent with other modelling exercises such as those by the

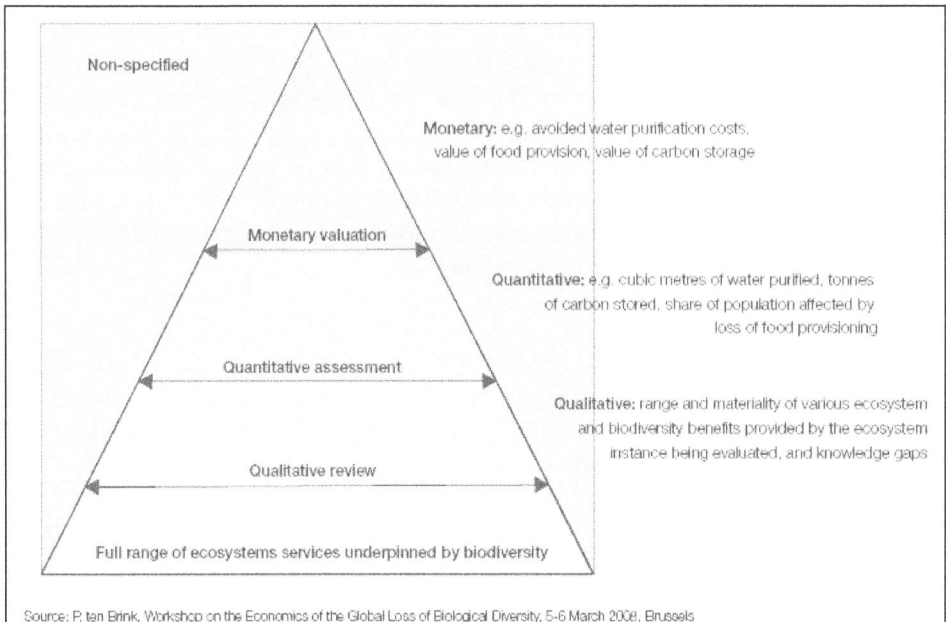

Source: P. ten Brink, Workshop on the Economics of the Global Loss of Biological Diversity, 5-6 March 2008, Brussels

Figure 2: Valuing ecosystem services.

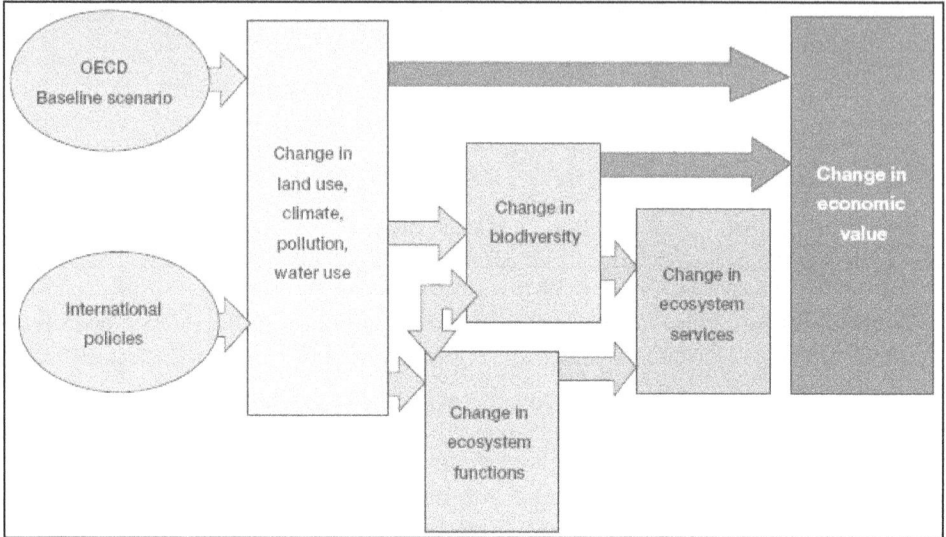

Figure 3: Establishing a scenario analysis.

FAO or other UN agencies. The model itself forecasts a slowing rate of biodiversity loss in Europe (compared to an increasing rate worldwide).

Assessing Changes in Ecosystem Services and Applying Monetary Values

Changes in land use and biodiversity are translated into changes in ecosystem services. The assessment relies to a large extent on the valuation literature, and creative solutions have been developed to extrapolate and fill data gaps.

The biggest difficulty has been to find studies to monetize changes in ecosystem services. While there are many case studies, not all regions, ecosystems and services are equally covered, and there were often difficulties in identifying values per hectare for use in such a widespread benefit transfer. Also, most studies are based on marginal losses, and the values are often location specific.

The Valuation Results

In the first years of the period 2000 to 2050, it is estimated that each year we are losing ecosystem services with a value equivalent to around EUR 50 billion from land-based ecosystems alone (it has to be noted that this is a welfare loss, not a GDP loss, as a large part of these benefits is currently not included in GDP). Losses of our natural capital stock are felt not only in the year of the loss, but continue over time, and are added to by losses in subsequent years of more biodiversity. These cumulative welfare losses could be equivalent to 7 per cent of annual consumption by 2050. This is a conservative estimate, because:

✯ It is partial, excluding numerous known loss categories, for example all marine biodiversity, deserts, the Arctic and Antarctic; some ecosystem services are excluded as well (disease regulation, pollination, ornamental services, etc.), while others are barely represented (*e.g.*, erosion control), or under-represented (*e.g.*, tourism); losses from invasive alien species are also excluded;

✯ Estimates for the rate of land-use change and biodiversity loss are globally quite conservative;

✯ The negative feedback effects of biodiversity and ecosystem loss on GDP growth are not fully accounted for in the model;

✯ Values do not account for non-linearities and threshold effects in ecosystem functioning.

Chapter 11
Global Warming

Global warming is the increase in the average temperature of the Earth's near-surface air and oceans since the mid-20th century and its projected continuation.

Global surface temperature increased $0.74 \pm 0.18°C$ ($1.33 \pm 0.32°F$) during the 100 years ending in 2005. The Intergovernmental Panel on Climate Change (IPCC) concludes that most of the temperature increase since the mid-twentieth century is "very likely" due to the increase in anthropogenic greenhouse gas concentrations. Natural phenomena such as solar variation and volcanoes probably had a small warming effect from pre-industrial times to 1950 and a small cooling effect from 1950 onward. These basic conclusions have been endorsed by at least 30 scientific societies and academies of science, including all of the national academies of science of the major industrialized countries. While individual scientists have voiced disagreement with these findings, the overwhelming majority of scientists working on climate change agree with the IPCC's main conclusions.

Climate model projections indicate that global surface temperature will likely rise a further 1.1 to $6.4°C$ (2.0 to $11.5°F$) during the twenty-first century. The uncertainty in this estimate arises from use of differing estimates of future greenhouse gas emissions and from use of models with differing climate sensitivity. Another uncertainty is how warming and related changes will vary from region to region around the globe. Although most studies focus on the period up to 2100, warming is expected to continue for more than a thousand years even if greenhouse gas levels are stabilized. This results from the large heat capacity of the oceans.

Increasing global temperature will cause sea levels to rise and will change the amount and pattern of precipitation, likely including an expanse of the subtropical desert regions. Other likely effects include increases in the intensity of extreme weather events, changes in agricultural yields, modifications of trade routes, glacier retreat, species extinctions and increases in the ranges of disease vectors.

Most national governments have signed and ratified the Kyoto Protocol aimed at reducing greenhouse gas emissions. Political and public debate continues regarding what, if any, action should be taken to reduce or reverse future warming or to adapt to its expected consequences.

Greenhouse Effect

The causes of the recent warming are an active field of research. The scientific consensus is that the increase in atmospheric greenhouse gases due to human activity caused most of the warming observed since the start of the industrial era, and the observed warming cannot be satisfactorily explained by natural causes alone. This attribution is clearest for the most recent 50 years, which is the period when most of the increase in greenhouse gas concentrations took place and for which the most complete measurements exist.

The greenhouse effect was discovered by Joseph Fourier in 1824 and first investigated quantitatively by Svante Arrhenius in 1896. It is the process by which absorption and emission of infrared radiation by atmospheric gases warm a planet's lower atmosphere and surface. Existence of the greenhouse effect as such is not disputed. The question is instead how the strength of the greenhouse effect changes when human activity increases the atmospheric concentrations of particular greenhouse gases.

Recent increases in atmospheric carbon dioxide (CO_2). The monthly CO_2 measurements display small seasonal oscillations in an overall yearly uptrend; each year's maximum is reached during the Northern Hemisphere's late spring, and declines during the Northern Hemisphere growing season as plants remove some CO_2 from the atmosphere.

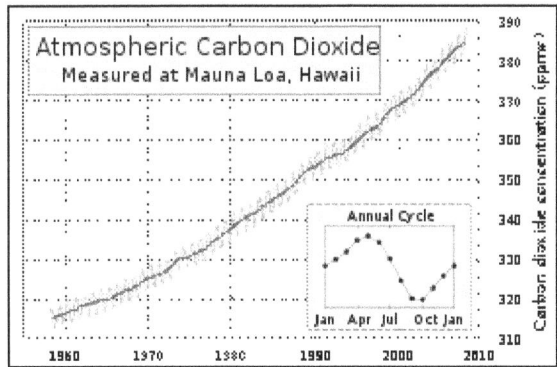

Naturally occurring greenhouse gases have a mean warming effect of about 33°C (59°F), without which Earth would be uninhabitable. On Earth the major greenhouse gases are water vapour, which causes about 36–70 per cent of the greenhouse effect (not including clouds); carbon dioxide (CO_2), which causes 9–26 per cent; methane (CH_4), which causes 4–9 per cent; and ozone, which causes 3–7 per cent.

Human activity since the industrial revolution has increased the atmospheric concentration of various greenhouse gases, leading to increased radiative forcing from CO_2, methane, tropospheric ozone, CFCs and nitrous oxide. The atmospheric concentrations of CO_2 and methane have increased by 36 per cent and 148 per cent respectively since the beginning of the industrial revolution in the mid-1700s. These levels are considerably higher than at any time during the last 650,000 years, the period for which reliable data has been extracted from ice cores. From less direct geological evidence it is believed that CO_2 values this high were last seen approximately 20 million years ago. Fossil fuel burning has produced approximately three-quarters of the increase in CO_2 from human activity over the past 20 years. Most of the rest is due to land-use change, in particular deforestation.

CO_2 concentrations are expected to continue to rise due to ongoing burning of fossil fuels and land-use change. The rate of rise will depend on uncertain economic, sociological, technological, and natural developments. The IPCC Special Report on Emissions Scenarios gives a wide range of future CO_2 scenarios, ranging from 541 to 970 ppm by the year 2100. Fossil fuel reserves are sufficient to reach this level and continue emissions past 2100 if coal, tar sands or methane clathrates are extensively exploited.

Some other hypotheses departing from the consensus view have been suggested to explain most of the temperature increase. One such hypothesis proposes that warming may be the result of variations in solar activity.

Solar Variation

A paper by Peter Stott and colleagues suggests that climate models overestimate the relative effect of greenhouse gases compared to solar forcing; they also suggest that the cooling effects of volcanic dust and sulfate aerosols have been underestimated. They nevertheless conclude that even with an enhanced climate sensitivity to solar forcing, most of the warming since the mid-

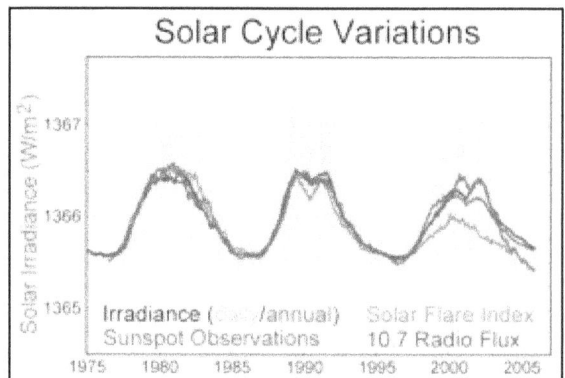

Solar variation over the last thirty years.

20th century is likely attributable to the increases in greenhouse gases. Another paper suggests that the Sun may have contributed about 45–50 per cent of the increase in the average global surface temperature over the period 1900–2000, and about 25–35 per cent between 1980 and 2000.

A different hypothesis is that variations in solar output, possibly amplified by cloud seeding via galactic cosmic rays, may have contributed to recent warming. It suggests magnetic activity of the sun is a crucial factor which deflects cosmic rays that may influence the generation of cloud condensation nuclei and thereby affect the climate.

One predicted effect of an increase in solar activity would be a warming of most of the stratosphere, whereas an increase in greenhouse gases should produce cooling there. The observed trend since at least 1960 has been a cooling of the lower stratosphere. Reduction of stratospheric ozone also has a cooling influence, but substantial ozone depletion did not occur until the late 1970s. Solar variation combined with changes in volcanic activity probably did have a warming effect from pre-industrial times to 1950, but a cooling effect since. In 2006, Peter Foukal and colleagues found no net increase of solar brightness over the last 1,000 years. Solar cycles led to a small increase of 0.07 per cent in brightness over the last 30 years. This effect is too small to contribute significantly to global warming. One paper by Mike Lockwood and Claus Fröhlich found no relation between global warming and solar radiation since 1985, whether through variations in solar output or variations in cosmic rays. Henrik Svensmark and Eigil Friis-Christensen, the main proponents of cloud seeding by galactic cosmic rays, disputed this criticism of their hypothesis. A 2007 paper found that in the last 20 years there has been no significant link between changes in cosmic rays coming to Earth and cloudiness and temperature.

Temperature Changes

Recent

Two millennia of mean surface temperatures according to different reconstructions, each smoothed on a decadal scale. The unsmoothed, annual value for 2004 is also plotted for reference.

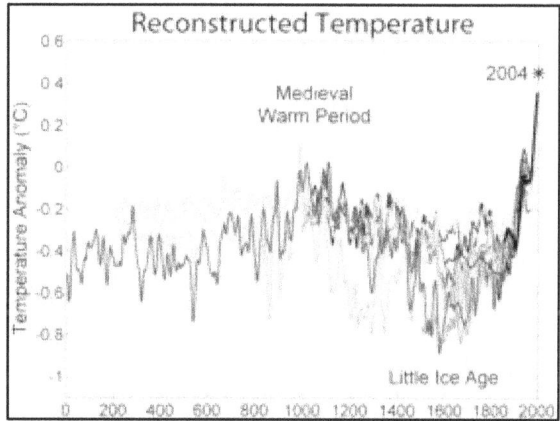

Global temperatures have increased by 0.75°C (1.35°F) relative to the period 1860–1900, according to the instrumental temperature record. This measured temperature increase is not significantly affected by the urban heat island effect. Since 1979, land temperatures have increased about twice as fast as ocean temperatures (0.25°C per decade against 0.13°C per decade). Temperatures in the lower troposphere have increased between 0.12 and 0.22 °C (0.22 and 0.4°F) per decade since 1979, according to satellite temperature measurements. Temperature is believed to have been relatively stable over the one or two thousand years before 1850, with possibly regional fluctuations such as the Medieval Warm Period or the Little Ice Age.

Sea temperatures increase more slowly than those on land both because of the larger effective heat capacity of the oceans and because the ocean can lose heat by evaporation more readily than the land. The Northern Hemisphere has more land than the Southern Hemisphere, so it warms faster. The Northern Hemisphere also has extensive areas of seasonal snow and sea-ice cover subject to the ice-albedo feedback. More greenhouse gases are emitted in the Northern than Southern Hemisphere, but this does not contribute to the difference in warming because the major greenhouse gases persist long enough to mix between hemispheres.

Based on estimates by NASA's Goddard Institute for Space Studies, 2005 was the warmest year since reliable, widespread instrumental measurements became available in the late 1800s, exceeding the previous record set in 1998 by a few hundredths of a degree. Estimates prepared by the World Meteorological Organization and the Climatic Research Unit concluded that 2005 was the second warmest year, behind 1998. Temperatures in 1998 were unusually warm because the strongest El Nino-Southern Oscillation in the past century occurred during that year.

Anthropogenic emissions of other pollutants–notably sulfate aerosols – can exert a cooling effect by increasing the reflection of incoming sunlight. This partially accounts for the cooling seen in the temperature record in the middle of the twentieth century, though the cooling may also be due in part to natural variability. James Hansen and colleagues have proposed that the effects of the products of fossil fuel combustion–CO_2 and aerosols – have largely offset one another, so that warming in recent decades has been driven mainly by non-CO_2 greenhouse gases.

Paleoclimatologist William Ruddiman has argued that human influence on the global climate began around 8,000 years ago with the start of forest clearing to provide land for agriculture and 5,000 years ago with the start of Asian rice irrigation. Ruddiman's interpretation of the historical record, with respect to the methane data, has been disputed.

Pre-Human Climate Variations

Curves of reconstructed temperature at two locations in Antarctica and a global record of variations in glacial ice volume. Today's date is on the left side of the graph.

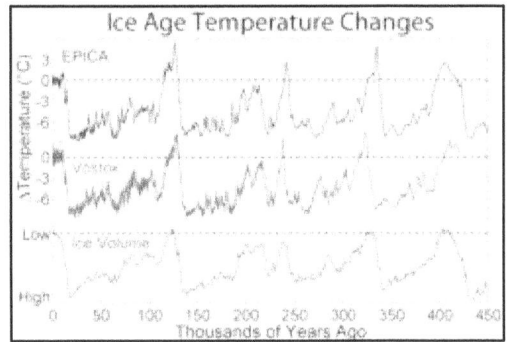
Ice Age Temperature Changes

Earth has experienced warming and cooling many times in the past. The recent Antarctic EPICA ice core spans 800,000 years, including eight glacial cycles timed by orbital variations with interglacial warm periods comparable to present temperatures.

A rapid buildup of greenhouse gases amplified warming in the early Jurassic period (about 180 million years ago), with average temperatures rising by 5°C (9°F). Research by the Open University indicates that the warming caused the rate of rock weathering to increase by 400 per cent. As such weathering locks away carbon in calcite and dolomite, CO_2 levels dropped back to normal over roughly the next 150,000 years.

Sudden releases of methane from clathrate compounds (the clathrate gun hypothesis) have been hypothesized as both a cause for and an effect of other warming events in the distant past, including the Permian–Triassic extinction event (about 251 million years ago) and the Paleocene–Eocene Thermal Maximum (about 55 million years ago).

Climate Models

Calculations of global warming prepared in or before 2001 from a range of climate models under the SRES A2 emissions scenario, which assumes no action is taken to reduce emissions.

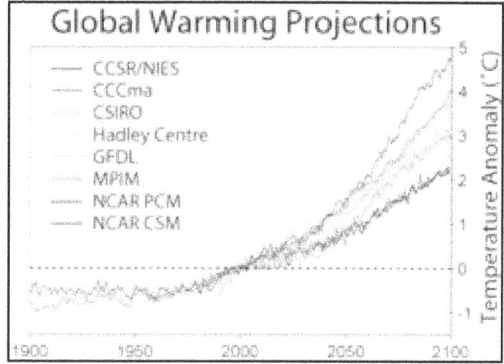

Global Warming Projections

CCSR/NIES
CCCma
CSIRO
Hadley Centre
GFDL
MPIM
NCAR PCM
NCAR CSM

The geographic distribution of surface warming during the 21st century calculated by the HadCM climate model if a business as usual scenario is assumed for economic growth and greenhouse gas emissions. In this figure, the globally averaged warming corresponds to 3.0°C (5.4°F).

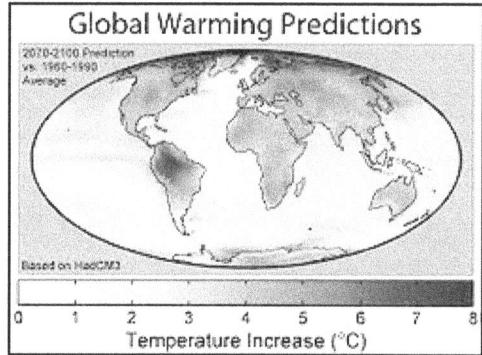

Global Warming Predictions

2070-2100 Prediction
vs 1960-1990
Average

Based on HadCM3

Temperature Increase (°C)

Scientists have studied global warming with computer models of the climate. These models are based on physical principles of fluid dynamics, radiative transfer, and other processes, with simplifications being necessary because of limitations in computer power and the complexity of the climate system. All modern climate models include an atmospheric model that is coupled to an ocean model and models for ice cover on land and sea. Some models also include treatments of chemical and biological processes. These models project a warmer climate due to increasing levels of greenhouse gases. However, even when the same assumptions of future greenhouse gas levels are used, there still remains a considerable range of climate sensitivity.

Including uncertainties in future greenhouse gas concentrations and climate modeling, the IPCC anticipates a warming of 1.1°C to 6.4°C (2.0°F to 11.5°F) by the end of the 21st century, relative to 1980–99. Models have also been used to help investigate the causes of recent climate change by comparing the observed changes to those that the models project from various natural and human-derived causes.

Current climate models produce a good match to observations of global temperature changes over the last century, but do not simulate all aspects of climate. These models do not unambiguously attribute the warming that occurred from approximately 1910 to 1945 to either natural variation or human effects; however, they suggest that the warming since 1975 is dominated by man-made greenhouse gas emissions.

Global climate model projections of future climate are forced by imposed greenhouse gas emission scenarios, most often from the IPCC Special Report on Emissions Scenarios (SRES). Less commonly, models may also include a simulation of the carbon cycle; this generally shows a positive feedback, though this response is uncertain (under the A2 SRES scenario, responses vary between an extra 20 and 200 ppm of CO_2). Some observational studies also show a positive feedback.

In May 2008, it was predicted that "global surface temperature may not increase over the next decade, as natural climate variations in the North Atlantic and tropical Pacific temporarily offset the projected anthropogenic warming", based on the inclusion of ocean temperature observations.

The representation of clouds is one of the main sources of uncertainty in present-generation models, though progress is being made on this problem.

A minor issue in climate modeling is the perceived mismatch between actual conditions and those projected by the models. A 2007 study by David Douglass and colleagues compared the composite output of 22 leading global climate models with actual climate data and found that the models did not accurately project observed changes to the temperature profile in the tropical troposphere. The authors note that their conclusions contrast strongly with those of recent publications based on essentially the same data. A 2008 paper published by a 17-member team led by Ben Santer of Lawrence Livermore National Laboratory noted serious mathematical flaws in the Douglass study, and found instead that deviations between the models and observations were statistically insignificant.

Economic Impacts of Global Warming

Many estimates of aggregate net economic costs of projected damages and benefits from climate change across the globe are now available. These are often expressed in terms of the social cost of carbon (SCC), the aggregate of future net benefits and costs, due to global warming from carbon dioxide emissions, that are discounted to the present. Peer-reviewed estimates of the SCC for 2005 have an average value of US$43 per tonne of carbon (tC) (*i.e.*, US$12 per tonne of carbon dioxide, tCO_2) but the range around this mean is large. For example, in a survey of 100 estimates, the values ran from US$-10 per tonne of carbon (US$-3 per tonne of carbon dioxide) up to US$350/tC (US$95 per tonne of carbon dioxide.)

One of the most widely noted projections on this issue is the *Stern Review*, a 2006 report by the former Chief Economist and Senior Vice-President of the World Bank Nicholas Stern, predicts that climate change will have a serious impact on economic growth without mitigation. The report suggests that an investment of one per cent of global GDP is required to mitigate the effects of climate change, with failure to do so risking a recession worth up to twenty per cent of global GDP. The *Stern Review* has been criticized by some economists, saying that Stern did not consider costs past 2200, that he used an incorrect discount rate in his calculations, and that stopping or significantly slowing climate change will require deep emission cuts everywhere. Other economists have supported Stern's approach, or argued that Stern's estimates are reasonable, even if the method by which he reached them is open to criticism. Research by Harvard Economist Martin Weitzman has suggested that structural uncertainty and low-probability high-impact risks are very important, and that "the influence on cost-benefit analysis of fat-tailed structural uncertainty about climate change, coupled with great unsureness about high-temperature damages, can outweigh the influence of discounting or anything else".

In the United States, insurance losses have also greatly increased. According to Choi and Fisher (2003) each 1 per cent increase in annual precipitation could enlarge catastrophe loss by as much as 2.8 per cent. Gross increases are mostly attributed to increased population and property values in vulnerable coastal areas, though there was also an increase in frequency of weather-related events like heavy rainfalls since the 1950s.

Infrastructure

Roads, airport runways, railway lines and pipelines, (including oil pipelines, sewers, water mains etc.) may require increased maintenance and renewal as they become subject to greater temperature variation and are exposed to weather that they were not designed for. Regions already adversely affected include areas of permafrost, which are subject to high levels of subsidence, resulting in buckling roads, sunken foundations, and severely cracked runways.

Investment

Venture capitalists and other investors have noted potential opportunities arising from global warming, as massive sums of money are needed for enhanced infrastructure as well as clean technologies that could help reduce emissions of global warming gases. AsJoel Makower, a noted expert on business and the environment, has pointed out, "For all the handwringing over the negative bottom-line impacts of climate change for most companies, a handful of large corporate interests may come out winners, creating potentially profitable opportunities for forward-thinking investors." These include companies investing in clean energy technologies such as solar energy and wind power, but also companies in other sectors: agriculture (to produce biofuels as well as biobased plastics that supplant petroleum-based ones), information technology companies (producing switches, routers, and software intended to create a more efficient, "smart grid", chemical companies (producing "green chemistry" alternatives to petrochemicals), and producers of more efficient motors for aircraft, automobiles, and industrial use.

Migration

Some Pacific Ocean island nations, such as Tuvalu, are concerned about the possibility of an eventual evacuation, as flood defense may become economically inviable for them. Tuvalu already has an ad hoc agreement with New Zealand to allow phased relocation.

In the 1990s a variety of estimates placed the number of environmental refugees at around 25 million. (Environmental refugees are not included in the official definition of refugees, which only includes migrants fleeing persecution.) The Intergovernmental Panel on Climate Change (IPCC), which advises the world's governments under the auspices of the UN, estimated that 150 million environmental refugees will exist in the year 2050, due mainly to the effects of coastal flooding, shoreline erosion and agricultural disruption (150 million means 1.5 per cent of 2050's predicted 10 billion world population).

Northwest Passage

Arctic ice thicknesses changes from 1950s to 2050s simulated in one of GFDL's R30 atmosphere-ocean general circulation model experiments

Melting Arctic ice may open the Northwest Passage in summer, which would cut 5,000 nautical miles (9,000 km) from shipping routes between Europe and Asia. This would be of particular relevance for supertankers which are too big to fit through the Panama Canal and currently have to go around the tip of South America. According to the Canadian Ice Service, the amount of ice in Canada's eastern Arctic Archipelago decreased by 15 per cent between 1969 and 2004.

While the reduction of summer ice in the Arctic may be a boon to shipping, this same phenomenon threatens the Arctic ecosystem, most notably polar bears which depend on ice floes. Subsistence hunters such as the Inuit peoples will find their livelihoods and cultures increasingly threatened as the ecosystem changes due to global warming.

Development

The combined effects of global warming may impact particularly harshly on people and countries without the resources to mitigate those effects. This may slow economic development and poverty reduction, and make it harder to achieve the Millennium Development Goals.

In October 2004 the Working Group on Climate Change and Development, a coalition of development and environment NGOs, issued a report Up in Smoke on the effects of climate change on development. This report, and the July 2005 report Africa–Up in Smoke? predicted increased hunger and disease due to decreased rainfall and severe weather events, particularly in Africa. These are likely to have severe impacts on development for those affected.

At the same time, in developing countries, the poorest often live on flood plains, because it is the only available space, or fertile agricultural land. These settlements often lack infrastructure such as dykes and early warning systems. Poorer communities also tend to lack the insurance, savings or access to credit needed to recover from disasters.

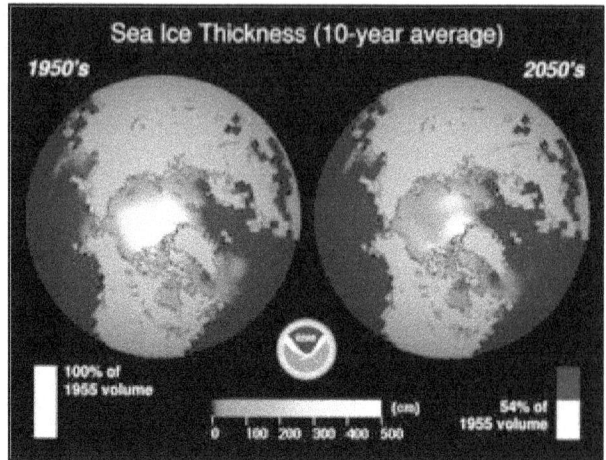

Environmental

Secondary evidence of global warming – reduced snow cover, rising sea levels, weather changes – provides examples of consequences of global warming that may influence not only human activities but also ecosystems. Increasing global temperature means that ecosystems may change; some species may be forced out of their habitats (possibly to extinction) because of changing conditions, while others may flourish. A 2004 study published in *Nature* estimates that between 15 and 37 per cent of known plant and animal species will be 'committed to extinction' by 2050. Few of the terrestrial ecoregions on Earth could expect to be unaffected.

Increasing carbon dioxide may increase ecosystems' productivity to a point. Ecosystems' unpredictable interactions with other aspects of climate change makes the possible environmental impact of this unclear, though. An increase in the total amount of biomass produced may not be necessarily positive: biodiversity can still decrease even though a relatively small number of species are flourishing.

Water Scarcity

Positive eustasy (sea-level rise) may contaminate groundwater, affecting drinking water and agriculture in coastal zones. Increased evaporation will reduce the effectiveness of reservoirs. Increased extreme weather means more water falls on hardened ground unable to absorb it, leading to flash floods instead of a replenishment of soil moisture or groundwater levels. In some areas, shrinking glaciers threaten the water supply. The availability of freshwater runoff from mountains for natural systems and human uses may also be impacted.

Higher temperatures will also increase the demand for water for the purposes of cooling and hydration.

In the Sahel, there has been on average a 25 per cent decrease in annual rainfall over the past 30 years.

Health

Direct Effects of Temperature Rise

Rising temperatures have two opposing direct effects on mortality: higher temperatures in winter reduce deaths from cold; higher temperatures in summer increase heat-related deaths.

The distribution of these changes obviously differs. Palutikof *et al.* calculate that in England and Wales for a 1°C temperature rise the reduced deaths from cold outweigh the increased deaths from heat, resulting in a reduction in annual average mortality of 7000. However, in the United States, only 1000 people die from the cold each year, while twice that number die from the heat. The 2006 United States heat wave has killed 139 people in California as of 29 July 2006. [Deaths of livestock have not been well-documented.] Fresno, in the central California valley, had six consecutive days of 110 degree-plus Fahrenheit temperatures.

The European heat wave of 2003 killed 22,000–35,000 people, based on normal mortality rates (Schär and Jendritzky, 2004). It can be said with 90 per cent confidence that past human influence on climate was responsible for at least half the risk of the 2003 European summer heat-wave (Stott *et al.*, 2004).

Spread of Disease

Global warming is expected to extend the favourable zones for vectors conveying infectious disease such as malaria. In poorer countries, this may simply lead to higher incidence of such diseases. In richer countries, where such diseases have been eliminated or kept in check by vaccination, draining swamps and using pesticides, the consequences may be felt more in economic than health terms, if greater spending on preventative measures is required.

Contamination by Sector and Cost of Reducing Fossil Fuel Use

Reducing greenhouse gas emissions depends in part on lowering consumption of fossil fuels. The key challenge is that nearly all forms of economic activity rely on fossil fuel energy sources, from transportation fuel, electricity from coal-fired plants, industrial furnaces to home and office heating. Reducing emissions can be achieved through gains in efficiency – producing the same benefits with smaller amounts of fossil energy, or by displacing fossil sources with non- or low-emitting sources. Low emission renewable energy sources such as wind, solar and biomass still represent only a small fraction of total energy consumption. The scale of current fossil energy dependence poses a substantial challenge. Gaining energy efficiency typically requires up-front investment, such as adding insulation, replacing energy-inefficient devices and processes, or buying hybrid vehicles. Some such investments can pay for themselves in the savings on energy bills, and the economic case for choosing them depends on the payback period. If an upgrade's payback is better than the risk-free interest rate, economic theory predicts individuals will choose the higher return of making the efficiency investment. If current pricing is not leading to this outcome, the cost of fossil energy is not yet high enough to drive adoption of available efficiency gains. (Social science researchers Kurani and Turrentine reported in 2004 that consumers often fail to make choices that have favorable payback period. They attribute the "uneconomic" choices to risk aversion, weighing potential losses much higher than potential gains.)

Advocates of mitigating climate change hold that greenhouse gas emissions must carry a price, so the market can internalize the externality of the impact of their emission. This could take the form of a carbon tax or of emission caps, with a market created for trading emission permits, much as was done in the USA for sulphate emissions blamed for acid rain. Thus the economic impact of avoiding greenhouse gas emissions depends on how much consumption will have to be avoided, and how quickly the economy can incorporate efficiency gains.

Some pundits have criticized such attempts at calculating the costs of mitigating climate change by avoiding fossil fuel consumption, pointing out that the opportunity costs of avoiding consumption are not (and cannot) be calculated and are likely to be more important than the expected benefits.

Many estimates of aggregate net economic costs of damages from climate change across the globe (*i.e.*, the social cost of carbon (SCC), expressed in terms of future net benefits and costs that are discounted to the present) are now available. Peer-reviewed estimates of the SCC for 2005 have an average value of US$43 per tonne of carbon (tC) (*i.e.*, US$12 per tonne of carbon dioxide) but the range around this mean is large, primarily due to the variation in discount rates used. For example, in a survey of 100 estimates, the values ran from US$-10 per tonne of carbon (US$-3 per tonne of carbon dioxide) up to US$350/tC (US$95 per tonne of carbon dioxide.

Mitigation and Adaptation

The costs of mitigating (reducing) global warming depend on a number of factors. One fundamental factor is the target level of atmospheric carbon dioxide: the lower the level, the sooner action must be taken if increases beyond the target level are to be avoided. The sooner action must be taken, the shorter the period over which costs must be spread, and the higher the absolute costs, as cheaper technologies which might emerge later are not yet available. A common target level (assumed by the United Kingdom) is 550 ppm (current levels are around 380 ppm, and rising at 2-3ppm per year). Signatories of the Kyoto Protocol committed themselves to targets that require lowering their national greenhouse gas emissions to a specified level relative to their actual 1990 emissions. Many nations set targets to reach a small percentage below 1990 levels, during the target period for Kyoto of 2008-12.

Another crucial factor in estimating the costs of climate change is the discount rate to apply. Normally a relatively high rate (*e.g.* 5 per cent -10 per cent) is applied, reflecting the cost of capital. However, where intergenerational issues involve potential irreversibilities such as climate change, a low discount rate (*e.g.* 1 per cent -4 per cent) may be applied. The difference is dramatic: at 4 per cent (a typical rate for social issues), avoiding $1m worth of climate change damage in 100 years' time is valued at nearly $20,000 today (net present value), whereas at 8 per cent it is valued at less than $500.

Another area for debate is the relationship between technological development and regulatory incentives: if regulation can induce substantial technological change, the costs of mitigation may be much lower.

Cost Estimates

IPCC TAR (Synthesis Report) suggested values of $78 bn to $1141 bn annual mitigation costs, amounting to 0.2 per cent to 3.5 per cent of current world GDP (which is around $35 trillion), or 0.3 per cent to 4.5 per cent of GDP if borne by the richest nations alone. As economic growth is expected to continue, the percentage would fall. In terms of cost per tonne of carbon emission avoided, the range (for a target of 550 ppm) is $18 to $80.

These cost estimates refer to reductions achieved through tradable emissions permits when those permits are given away to polluters. If the reductions are achieved through emission taxes or auctioned permits, and the revenue is used to reduce distortionary taxes, the TAR III synthesis report concludes that "[depending] on the existing tax structure, type of tax cuts, labour market conditions and method of recycling. it is possible that the economic benefits may exceed the costs of mitigation." Nordhaus and Boyer calculated that the present value cost of the Kyoto Protocol would be $800 billion to $1,500 billion if implemented as efficiently as possible. Richard Tol estimates that the net present value cost to be more than $2.5 trillion.

Azar and Schneider (2002) observe that global output in 1990 was around $20 trillion. If it grew steadily at 2.1 per cent per annum it would be just short of $200 trillion by 2100. They thereby make the point that the calculated present value costs of mitigation would look smaller if scaled against 2100 output than if scaled against 1990 output. However, neither comparator is relevant to the question of whether the likely benefits from mitigation exceed the costs.

A 2008 study, not peer-reviewed, by the consulting company McKinsey Global Institute uses cost curve analysis to estimate that it is possible to stabilize global greenhouse gas concentrations at 450 to 500 ppm CO_2-e with macroeconomic costs in the order of 0.6-1.4 per cent of global GDP by 2030.

Lord Peter Levene, chairman of Lloyd's of London, said on 12 April 2007 that the threat of climate change must be an integral part of every company's risk analysis.

Benefits

Nordhaus and Boyer estimated that the present value of benefits from mitigation under the Kyoto Protocol would be $120 billion, far below the likely costs. "Other studies reach similar conclusions". Richard Tol concludes that "the emissions targets agreed in the Kyoto Protocol are irreconcilable with economic rationality."

However, the Stern Review produced much larger benefit estimates, of between 5 per cent and 20 per cent of GDP. The difference reflected a number of factors, the most important of which were the choice of discount rate, the use of welfare weighting for effects on people in poor countries, a greater weight on damage to the natural environment and the use of more up-to-date scientific estimates of likely damage.

In addition to avoiding the costs of the business-as-usual scenario, mitigation actions can bring other benefits, depending on factors such as the technology used. These include, for example, the reduced economic impact from oil supply disruptions and/or price rises, if mitigation reduces oil dependence. This may be of particular benefit to non-oil-exporting developing countries, which suffer greater economic impact from oil price rises. Co-benefits from ending deforestation include protection of biodiversity, benefits for indigenous people, research and development possibilities, tourism, and some protection from extreme weather events. (Stern Review, p. 280).

Optimal Strategies for Mitigation

Financial and technological strategies can have a major impact on reaching a particular target atmospheric greenhouse gas concentration.

   ☆ Carbon tax
   ☆ Carbon emissions trading
   ☆ A hybrid between a carbon tax and an emissions trading scheme, this can be thought of as an emissions trading scheme with a price cap, a price floor, or both. A price cap can be realized by governments being able to sell an unlimited amount extra permits at a given price (the price of the cap). A

price floor can be realized by governments buying back permits if the price goes below the value of the floor, or by emitters paying a fee when they exercise the permit (so the effective carbon price is equal to the sum of the permit price and the exercise fee).

☆ Regulation

☆ Reducing the carbon intensity of energy via Nuclear power or Renewable Energy

☆ Energy efficiency

"No regrets" policies – notably reducing fossil fuel subsidies, which is predicted to increase growth whilst reducing CO_2 emissions. Article 2 of the Kyoto Protocol specifies a progressive removal of subsidies and reform of taxes as a means of achieving reduction commitments.

McKibbin and Wilcoxen argue that a combination of long term carbon price signals and short terms caps on economic cost is needed to address both economic efficiency, equity sharing and political feasibility.

The Stern Review recommends adopting a quantative global stabilisation target range for the stock of greenhouse gases as a foundation for policy. It suggests that this target range would be likely to be somewhere between 450-550 ppm CO_2-e. It also recommends a carbon price signal through the use of a carbon tax or emissions trading scheme.

Brink et al (2005) showed that the costs of mitigation can be reduced by considering the inter-relationships of different greenhouse gas, and the differential impact that different technological decisions may have on their emissions.

Chapter 12

Acid Rain

Acid rain is precipitation containing higher than normal amounts of nitric and sulfuric acid. According to the EPA, acid rain can either occur by wet deposition or dry deposition, occurring when sulfur dioxide and nitrogen oxide react with the atmosphere. During wet deposition the gases react with water, oxygen and other chemicals. When this happens, various acidic compounds form, most commonly sulfuric and nitric acid. These acids are blown by the wind into areas with wet weather, where they fall in the form of rain or snow. If the acids are blown to areas with dry weather, they fall as dry deposition, clinging to dust or smoke, sticking to buildings and trees. The acid is then washed into the water system by rain.

Acid rain forms from both natural and man-made sources. Volcanoes and decaying vegetation can form sulfur dioxide and nitrogen oxides. Emissions from burning fossil fuels also produce acid rain. According to the EPA, in the United States most of the sulfur dioxide and a quarter of nitrogen oxides "come from electric power generation that relies on burning fossil fuels, like coal." Winds have been known to carry the acid rain into parts of Canada.

Acid Rain

Acid rain occurs when sulfur dioxide and nitrogen oxides are emitted into the atmosphere, undergo chemical transformations and are absorbed by water droplets in clouds. The droplets then fall to earth as rain, snow, or sleet. This can increase the acidity of the soil, and affect the chemical balance of lakes and streams. Acid rain is sometimes used more generally to include all forms of acid deposition – both wet deposition, where acidic gases and particles are removed by rain or other precipitation, and dry deposition removal of gases and particles to the Earth's surface in the absence of precipitation.

Acid rain is defined as any type of precipitation with a pH that is unusually low (Brimblecombe, 1996). Dissolved carbon dioxide dissociates to form weak carbonic acid giving a pH of approximately 5.6 at typical atmospheric concentrations of CO_2 (Seinfeld and Pandis, 1998). Therefore a pH of <5.6 has sometimes been used as a definition of acid rain. However, natural sources of acidity mean that in remote areas rain has a pH which is between 4.5 and 5.6 with an average value of 5.0 and so rain with a pH <5 is a more appropriate definition.

Acid rain accelerates weathering in carbonate rocks and accelerates building weathering. It also contributes to acidic rivers, streams, and damage to trees at high elevation. Efforts to combat this phenomenon are ongoing.

History and Trends

Acid rain was first reported in Manchester, England, which was an important city during the Industrial Revolution. In 1852, Robert Angus Smith found the relationship between acid rain and atmospheric pollution. The term "acid rain" was used for the first time by him in 1872 (Seinfeld and Pandis, 1998).

Though acid rain was discovered in 1852, it wasn't until the late 1960s that scientists began widely observing and studying the phenomenon. Canadian Harold Harvey was among the first to research a "dead" lake. Public awareness of acid rain in the U.S increased in the 1990s after the New York Times promulgated reports from the Hubbard Brook Experimental Forest in New Hampshire of the myriad deleterious environmental effects demonstrated to result from it.

Evidence for an increase in the levels of acid rain comes from analysing layers of glacial ice. These show a sudden decrease in pH from the start of the industrial revolution of 6 to 4.5 or 4. Other information has been gathered from studying organisms known as diatoms which inhabit ponds. Over the years these die and are deposited in layers of sediment on the bottoms of the ponds. Diatoms thrive in certain pHs, so the numbers of diatoms found in layers of increasing depth give an indication of the change in pH over the years.

Since the industrial revolution, emissions of sulfur and nitrogen oxides to the atmosphere have increased. Industrial and energy-generating facilities that burn fossil fuels, primarily coal, are the principal sources of increased sulfur oxides. Occasional pH readings of well below 2.4 (the acidity of vinegar) have been reported in industrialized areas. These sources, plus the transportation sector, are the major originators of increased nitrogen oxides.

The problem of acid rain not only has increased with population and industrial growth, but has become more widespread. The use of tall smokestacks to reduce local pollution has contributed to the spread of acid rain by releasing gases into regional atmospheric circulation. Often deposition occurs a considerable distance from its formation, with mountainous regions tending to receive the most (simply because of their higher rainfall). An example of this effect is the frequent low pH of rain which falls in Scandinavia compared to the local emissions.

Industrial acid rain is a substantial problem in China, Eastern Europe, Russia and areas down-wind from them. Acid rain from power plants in the midwest United

States has also harmed the forests of upstate New York and New England. These areas all burn sulphur-containing coal to generate heat and electricity.

Emissions of Chemicals Leading to Acidification

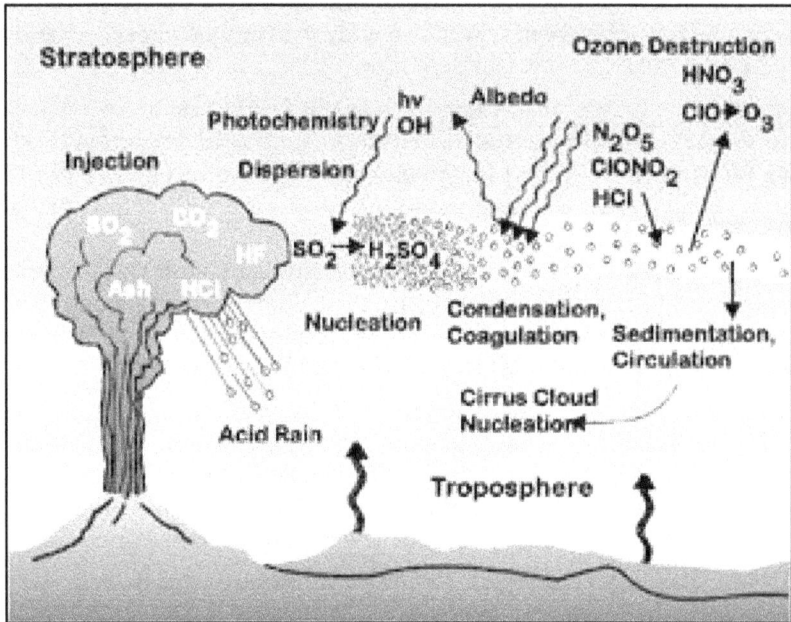

Figure 1

The most important gas which leads to acidification is sulfur dioxide. Emissions of nitrogen oxides which are oxidised to form Nitric acid are of increasing importance due to stricter controls on emissions of sulfur containing compounds. 70 Tg(S) per year in the form of SO_2 comes from fossil fuel combustion and industry, 2.8 Tg(S) from wildfires, and 7-8 Tg(S) per year from volcanoes.

Natural Emissions

The principal natural phenomena that contribute acid-producing gases to the atmosphere are emissions from volcanoes and those from biological processes that occur on the land, in wetlands, and in the oceans. The major biological source of sulfur containing compounds is Dimethyl sulphide.

The effects of acidic deposits have been detected in glacial ice thousands of years old in remote parts of the globe.

Human Emissions

The principal cause of acid rain is sulfur and nitrogen compounds from human sources, such as electricity generation and motor vehicles. The gases can be carried hundreds of miles in the atmosphere before they are converted to acids and deposited.

Gas Phase Chemistry

In the gas phase sulphur dioxide is oxidised by reaction with the hydroxyl radical via a termolecular reaction:

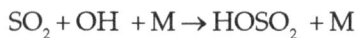

$$SO_2 + OH + M \rightarrow HOSO_2 + M$$

which is followed by:

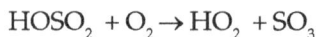

$$HOSO_2 + O_2 \rightarrow HO_2 + SO_3$$

In the presence of water sulfur trioxide is converted rapidly to sulphuric acid:

$$SO_3 + H_2O + M \rightarrow H_2SO_4 + M$$

Nitric acid is formed by the reaction of OH with Nitrogen dioxide:

$$NO_2 + OH + M \rightarrow HNO_3 + M$$

Chemistry in Cloud Droplets

When clouds are present the loss rate of SO_2 is faster than can be explained by gas phase chemistry alone. This is due to reactions in the liquid water droplets.

Hydrolysis

Sulphur dioxide dissolves in water and then, like carbon dioxide, hydrolyses in a series of equilibrium reactions:

$$SO_2 \, (g)^+ H_2O \rightarrow SO_2 \, H_2O \, SO_2 \, H_2O \rightarrow H^{++}HSO_3^- \, HSO_3^- \rightarrow H^{++}SO_3^{2-}$$

Oxidation

There are a large number of aqueous reactions of sulphur which oxidise it from S(IV) to S(VI) leading to the formation of sulfuric acid. The most important oxidation reactions are with ozone, hydrogen peroxide and oxygen (reactions with oxygen are catalysed by Iron and Manganese in the cloud droplets).

For more information see Seinfeld and Pandis (1998).

Aerosol Formation

In the gas phase sulfuric and nitric can condense on existing aerosols or nucleate to form new aerosols. The nucleation process is an important source of new particles in the atmosphere and so emissions of sulfur containing compounds, as well as causing acidification also have a climate effect.

Acid Deposition

Processes involved in acid deposition (note that only SO_2 and NO_x play a significant role in acid rain).

Wet Deposition

Wet deposition of acids occurs when any form of precipitation (rain, snow, etc.) removes acids from the atmosphere and delivers it to the Earth's surface. This can result from the deposition of acids produced in the raindrops (see aqueous phase

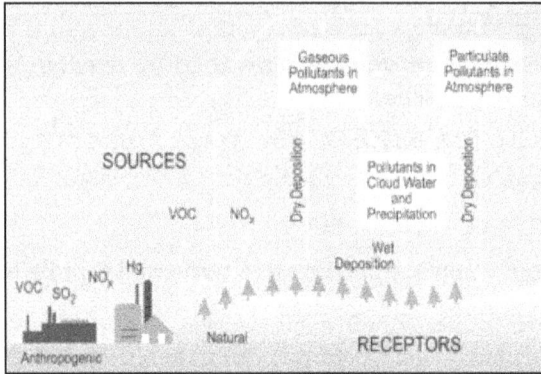

Figure 2

chemistry above) or by the precipitation removing the acids either in clouds or below clouds. Wet removal of both gases and aerosol are both of importance for wet deposition.

Dry Deposition

Acid deposition also occurs via dry deposition in the absence of precipitation. This can be responsible for as much as 20 to 60 per cent of total acid deposition. This occurs when particles and gases stick to the ground, plants or other surfaces.

Adverse Effects

Decades of enhanced acid input has increased the environmental stress on high elevation forests and aquatic organisms in sensitive ecosystems. In extreme cases, it has altered entire biological communities and eliminated some fish species from certain lakes and streams. In many other cases, the changes have been more subtle, leading to a reduction in the diversity of organisms in an ecosystem. This is particularly true in the northeastern United States, where the rain tends to be most acidic, and often the soil has less capacity to neutralize the acidity.

Acid rain also can damage certain building materials and historical monuments.

Some scientists have suggested links to human health, but none have been proven.

Effects on Lake Ecology

There is a strong relationship between lower pH values and the loss of populations of fish in lakes. Below 4.5 virtually no fish survive, whereas levels of 6 or higher promote healthy populations. Acid in water inhibits the production of enzymes which enable fish's larvae to escape their eggs. It also mobilizes toxic metals such as aluminium in lakes. Aluminium causes some fish to produce an excess of mucus around their gills, preventing proper ventilation. Phytoplankton growth is inhibited by high acid levels, and animals which feed on it suffer.

Many lakes are subject to natural acid runoff from acid soils, and this can be triggered by particular rainfall patterns that concentrate the acid. An acid lake with newly-dead fish is not necessarily evidence of severe air-pollution.

Effects of Acid Rain on Soil Biology

Soil biology can be seriously damaged by acid rain. Some tropical microbes can quickly consume acids (Rodhe, 2005) but other types of microbe are unable to tolerate low pHs and are killed. The enzymes of these microbes are denatured (changed in shape so they no longer function) by the acid.

The hydronium ions of acid rain also mobilize toxins and leach away essential nutrients.

Forest soils tend to be inhabited by fungi, but acid rain shifts forest soils to be more bacterially dominated. In order to fix nitrogen many trees rely on fungi in a symbiotic relationship with their roots. If acidity inhibits the growth of these mycorrhizae associations this could lead to trees struggling to fix nitrogen without their symbiotic partners.

Other Adverse Effects

Trees are harmed by acid rain in a variety of ways. The waxy surface of leaves is broken down and nutrients are lost, making trees more susceptible to frost, fungi, and insects. Root growth slows and as a result fewer nutrients are taken up. Toxic ions are mobilized in the soil, and valuable minerals are leached away or (as in the case of phosphate) become bound to aluminium or iron compounds, or to clay.

The toxic ions released due to acid rain form the greatest threat to humans. Mobilized copper has been implicated in outbreaks of diarrhea/diarrhoea in young children and it is thought that water supplies contaminated with aluminium cause Alzheimer's disease.

Acid rain can cause corrosion of ancient and valuable statues and has caused considerable damage. This is because the sulfuric acid in the rain chemically reacts with the calcium in the stones (lime stone, sandstone, marble and granite) to create gypsum, which then flakes off. This is also commonly seen on old gravestones where the acid rain can cause the inscription to become completely illegible.

Acid rain also causes an increased rate of oxidation for iron.

Prevention Methods

Technical Solutions

In the United States, many coal-burning power plants use Flue gas desulfurization (FGD) to remove sulfur-containing gases from their stack gases. An example of FGD is the wet scrubber which is commonly used in the U.S. and many other countries. A wet scrubber is basically a reaction tower equipped with a fan that extracts hot smoky stack gases from a power plant into the tower. Lime or limestone in slurry form is also injected into the tower to mix with the stack gases and combine with the sulfur dioxide present. The calcium carbonate of the limestone produces pH-neutral calcium sulfate that is physically removed from the scrubber. That is, the scrubber turns sulfur pollution into industrial sulfates.

In some areas the sulfates are sold to chemical companies as gypsum when the purity of calcium sulfate is high. In others, they are placed in a land-fill.

International Treaties

A number of international treaties on the long range transport of atmospheric pollutants have been agreed, *e.g.*, Sulphur Emissions Reduction Protocol and Convention on Long-Range Transboundary Air Pollution.

Emissions Trading

An even more benign regulatory scheme involves emission trading. In this scheme, every current polluting facility is given an emissions license that becomes part of capital equipment. Operators can then install pollution control equipment, and sell parts of their emissions licenses. The main effect of this is to give operators real economic incentives to install pollution controls. Since public interest groups can retire the licenses by purchasing them, the net result is a continuously decreasing and more diffused set of pollution sources. At the same time, no particular operator is ever forced to spend money without a return of value from commercial sale of assets.

Acid rain causes acidification of lakes and streams and contributes to the damage of trees at high elevations (for example, red spruce trees above 2,000 feet) and many sensitive forest soils. In addition, acid rain accelerates the decay of building materials and paints, including irreplaceable buildings, statues, and sculptures that are part of our nation's cultural heritage. Prior to falling to the earth, sulfur dioxide (SO_2) and nitrogen oxide (NO_x) gases and their particulate matter derivatives – sulphates and nitrates – contribute to visibility degradation and harm public health.

Over the years, scientists, foresters, and others have watched some forests grow more slowly without knowing why. The trees in these forests do not grow as quickly as usual. Leaves and needles turn brown and fall off when they should be green and healthy.

Researchers suspect that acid rain may cause the slower growth of these forests. But acid rain is not the only cause of such conditions. Other air pollutants, insects, diseases and drought are some other causes that harm plants. Also, some areas that receive acid rain show a lot of damage, while other areas that receive about the same amount of acid rain do not appear to be harmed at all. However, after many years of collecting information on the chemistry and biology of forests, researchers are beginning to understand how acid rain works on the forest soil, trees, and other plants.

Acid Rain on the Forest Floor

A spring shower in the forest washes leaves and falls through the trees to the forest floor below. Some of the water soaks into the soil. Some trickles over the ground and runs into a stream, river or lake. That soil may neutralize some or all of the acidity of the acid rainwater. This ability of the soil to resist some pH change is called buffering capacity. A buffer resists changes in pH. Without buffering capacity, soil pH would change rapidly. Midwestern states like Nebraska and Indiana have soils that are well buffered. Places in the mountainous northeast, like New York's Adirondack Mountains, have soils that are less able to buffer acids. Since there are many natural sources of acid in forest soils, soils in these areas are more susceptible to effects from acid rain.

How Acid Rain Harms Trees

Acid rain does not usually kill trees directly. Instead, it is more likely to weaken the trees by damaging their leaves, limiting the nutrients available to them, or poisoning them with toxic substances slowly released from the soil.

Scientists believe that acidic water dissolves the nutrients and helpful minerals in the soil and then washes them away before the trees and other plants can use them to grow. At the same time, the acid rain causes the release of toxic substances such as aluminum into the soil. These are very harmful to trees and plants, even if contact is limited. Toxic substances also wash away in the runoff that carries the substances into streams, rivers, and lakes. Less of these toxic substances are released when the rainfall is cleaner.

Even if the soil is well buffered, there can be damage from acid rain. Forests in high mountain regions receive additional acid from the acidic clouds and fog that often surround them. These clouds and fog are often more acidic than rainfall. When leaves are frequently bathed in this acid fog, their protective waxy coating can wear away. The loss of the coating damages the leaves and creates brown spots. Leaves turn the energy in sunlight into food for growth. This process is called photosynthesis. When leaves are damaged, they cannot produce enough food energy for the tree to remain healthy.

Once trees are weak, they can be more easily attacked by diseases or insects that ultimately kill them. Weakened trees may also become injured more easily by cold weather.

Acid rain can harm other plants in the same way it harms trees. Food crops are not usually seriously affected, however, because farmers frequently add fertilizers to the soil to replace nutrients washed away. They may also add crushed limestone to the soil. Limestone is a basic material and increases the ability of the soil to act as a buffer against acidity.

Water

The effects of acid rain are most clearly seen in the aquatic, or water, environments, such as streams, lakes, and marshes. Acid rain flows to streams, lakes, and marshes after falling on forests, fields, buildings, and roads. Acid rain also falls directly on aquatic habitats.

Most lakes and streams have a pH between 6 and 8. However, some lakes are naturally acidic even without the effects of acid rain. Lakes and streams become acidic (pH value goes down) when the water itself and its surrounding soil cannot buffer the acid rain enough to neutralize it. In areas like the Northeastern United States where soil buffering is poor, some lakes now have a pH value of less than 5. One of the most acidic lakes reported is Little Echo Pond in Franklin, New York. Little Echo Pond has a pH of 4.2. Lakes and streams in the western United States are usually not acidic. Because of differences in emissions and wind patterns, levels of acid deposition are generally lower in the western United States than in the eastern United States.

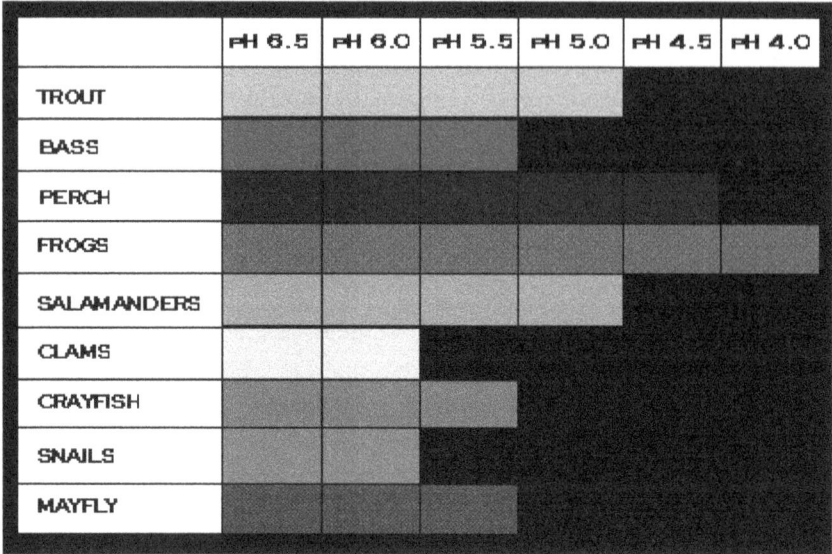

	pH 6.5	pH 6.0	pH 5.5	pH 5.0	pH 4.5	pH 4.0
TROUT						
BASS						
PERCH						
FROGS						
SALAMANDERS						
CLAMS						
CRAYFISH						
SNAILS						
MAYFLY						

Figure 3

This chart shows that not all fish, shellfish, or their food insects can tolerate the same amount of acid:

Generally, the young of most species are more sensitive than adults. Frogs may tolerate relatively high levels of acidity, but if they eat insects like the mayfly, they may be affected because part of their food supply may disappear. As lakes and streams become more acidic, the numbers and types of fish and other aquatic plants and animals that live in these waters decrease. Some types of plants and animals are able to tolerate acidic waters. Others, however, are acid-sensitive and will be lost as the pH declines. Some acid lakes have no fish. At pH 5, most fish eggs cannot hatch. At lower pH levels, some adult fish die. Toxic substances like aluminum that wash into the water from the soil may also kill fish.

Together, biological organisms and the environment in which they live are called an ecosystem. The plants and animals living within an ecosystem are highly interdependent. For example, fish eat other fish and also other plants and animals that live in the lake or stream. If acid rain causes the loss of acid-sensitive plants and animals, then fish that rely on these organisms for food may also be affected.

Human-Made Materials

Acid rain eats away at stone, metal, paint – almost any material exposed to the weather for a long period of time. Human-made materials gradually deteriorate even when exposed to unpolluted rain, but acid rain accelerates the process. Acid rain can cause marble statues carved long ago to lose their features. Acid rain has the same effect on buildings and monuments. Repairing acid rain damage to houses, buildings, and monuments can cost billions of dollars. Ancient monuments and buildings, such as the Parthenon in Greece, can never be replaced.

Here is a way for you to observe the effect of acid rain on marble and limestone, two building materials commonly used in monuments, ancient buildings, and in many modern structures.

☆ Place a piece of chalk in a bowl with white vinegar.

☆ Place another piece in a bowl of tap water.

☆ Leave the dishes overnight.

The next day, see if you can see which piece of chalk is more worn away.

This experiment with chalk allows you to see the effect of acid rain on marble and limestone because chalk is made of calcium carbonate, a compound occurring in rocks, such as marble and limestone, and in animal bones, shells, and teeth.

Effects of Acid Rain on People

Acid rain looks, feels, and tastes just like clean rain. The harm to people from acid rain is not direct. Walking in acid rain, or even swimming in an acid lake, is no more dangerous than walking or swimming in clean water. The air pollution that causes acid rain is more damaging to human health. Sulfur dioxide and nitrogen oxides, the major sources of acid rain, can irritate or even damage our lungs.

The pollutants that cause acid rain can also reduce visibility – limiting how far into the distance we can see.

The primary pollutants associated with acid rain and poor visibility are human-made sulfur dioxide emissions. These emisisons form small sulfate particles, or aerosols, in the atmosphere. These aerosols reduce visibility by scattering light. Sulphate aerosols are the main cause of poor visibility in the eastern United States.

Nitrogen oxide emissions are also associated with the acid rain problem. They, too, can form aerosols in the atmosphere that significantly reduce visibility. Nitrate aerosols are often the main cause for poor visibility in the western United States where sulfur dioxide emissions and humidity are lower than in the east.

Since the beginning of the Industrial Revolution soiling and degradation of buildings in urban areas has been noticeable. The cause of this has often been attributed to the effects of air pollution. The pollutants that form acid rain are principally sulphur dioxide and nitrogen oxides; both of these are released from the combustion of fossil fuels like coal and oil. Since the Industrial Revolution emissions of both have increased. UK Sulphur dioxide (SO_2) emissions peaked in the 1960s but have since declined by over 80 per cent. In 1999 emissions of sulphur dioxide were approximately 1.2 million tonnes. Emissions of nitric oxides and nitrogen dioxides, collectively known as NO_x, have fallen since 1990; emissions in 1999 were around 1.6 million tonnes.

Despite the reduction in emissions there is no clear evidence that cleaner air has brought about a reduction in building degradation. In fact, buildings that have withstood thousands of years of weathering have in the last 25 years or so begun to deteriorate rapidly. This can be attributed to the permanent alteration of stone surfaces by sulphation, a process whereby the exposed surface of limestone dissolves away as rainfall washes away the sulphated layers.

It is only in the last decade or so that attempts have been made to quantify the amount of damage that has been caused to materials as a result of acid deposition. Concern about the effects of acid rain on building materials was raised in a House of Commons Select Committee report in September 1984. As part of the governments response, the Buildings Effect Review Group (BERG) was established to give considered advice on the effects of acid deposition on buildings. It is only relatively recently that the spatial concentrations of acid rain pollutants and their transport mechanisms have become fully understood so more accurate estimates of the damage that may occur to buildings can be made.

Materials Affected

The list of materials affected by acid deposition is very long as most materials are liable to some degree of damage. Those most vulnerable are: limestone; marble; carbon-steel; zinc; nickel; paint and some plastics. Stone decay can take several forms, including the removal of detail from carved stone, and the build-up of black gypsum crusts in sheltered areas. Metal corrosion is caused primarily by oxygen and moisture, although SO_2 does accelerate the process. Most structures and buildings are affected by acid deposition to some degree because few materials are safe from these effects. In addition to atmospheric attack structures that are submerged in acidified waters such as foundations and pipes can also be corroded.

The Chemistry of Corrosion

Wet and dry deposition both contribute to the corrosion of materials. Dry deposition consists of gaseous and particulate matter that falls to Earth close to the source of emissions causing direct damage. Sulphur dioxide often falls as dry deposition within 30 km of its source. Wet deposition occurs when the pollutants are spread high into the atmosphere, where they react with water vapour in clouds to form dilute acids. The effects are felt much further afield and therefore wet deposition can affect areas that are many tens of kilometres away from any sources of pollution.

Calcium carbonate in certain stones dissolves in dilute sulphuric acid to form calcium sulphate:

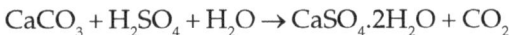

$$CaCO_3 + H_2SO_4 + H_2O \rightarrow CaSO_4.2H_2O + CO_2$$

This has two effects. Firstly it causes the surface of the stone to break up; secondly, a black skin of gypsum (calcium sulphate) forms which blisters off exposing more stone. When the gypsum crystals form they can grow into the stone, and the process may continue for up to 50 years. This is known as the Memory Effect.

Sulphur dioxide is the main pollutant in respect to corrosion but others also take their toll including NO_x, carbon dioxide (CO_2), ozone (on organic materials) and sea salt from sea spray. Research has revealed that when nitrogen dioxide (NO_2) is present with SO_2, increased corrosion rates occur. This is because the NO_2 oxidises the SO_2 to sulphite (SO_3) thereby promoting further SO_2 absorption. The Review Group on Acid Rain report in 1990 indicated that in remote areas wet deposition will predominate, whereas in Eastern England dry deposition will predominate. This finding is supported by a study of south-east England, which suggests that up to 40 per cent of total damage is due to dry deposition.

The interactions between materials and pollutants are very complex and many variables are involved. Deposition of pollutants onto surfaces depends on atmospheric concentrations of the pollutants and the climate and micro-climate around the surface. Once the pollutants are on the surface, interactions will vary depending on the amount of exposure, the reactivity of different materials and the amount of moisture present. The last factor is particularly important because the SO_2 that falls as dry deposition is oxidised to sulphuric acid in the presence of moisture on the surface.

Studies Undertaken

There are a number of studies that have been initiated looking at the effects of acid deposition on different materials. The National Materials Exposure Programme (NMEP) was initiated by BERG in 1987 and consists of 29 sites throughout the country. Samples of different materials are exposed at the sites for a period of not less than four years, during which time data on meteorological conditions and atmospheric conditions will be collected and corrosion rates monitored. The UK NMEP is also part of the International Materials Exposure Programme set up under the United Nations Economic Commission for Europe framework, in which materials are exposed to polluted environments in Europe and North America.

Examples of Damage

The effects of acid deposition on modern buildings are considerably less damaging than the effects on ancient monuments. Limestone and calcareous stones which are used in most heritage buildings are the most vulnerable to corrosion and need continued renovation.

Evidence of the damaging effect of acid deposition can be seen throughout the world. For example, world famous structures as the Taj Mahal, Cologne Cathedral, Notre Dame, the Colosseum and Westminster Abbey have all been affected.

Acid Formation in the Atmosphere

First, let us review some basic chemistry as it applies to acid precipitation.

Carbonic acid forms naturally in the atmosphere due to the reaction of water (H2O) and carbon dioxide (CO_2),

$$H_2O + CO_2 \rightarrow H_2CO_3$$

while the burning of coal and other organics adds sulfur dioxide (SO_2) and Nitrous oxides (NO_x) to the atmosphere where they react to form sulfuric acid and nitric acid,

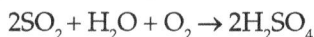

$$2SO_2 + H_2O + O_2 \rightarrow 2H_2SO_4$$
$$4NO_2 + 2H_2O + O_2 \rightarrow 4HNO_3$$

All of these acids will be buffered by reacting with rocks, minerals, etc. on the earth's surface. The most important (and fastest) buffering comes from the reaction with (weathering of) calcite in the form of limestone, dolomite or marble.

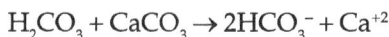

$$H_2CO_3 + CaCO_3 \rightarrow 2HCO_3^- + Ca^{+2}$$

When this reaction occurs, the acid is neutralized and the calcite dissolved. While the reaction with calcite is very fast (the standard test for calcite in introductory geology labs is to put very dilute acid on a sample to see if it bubbles (reacts)), the reaction with other rocks is very slow, so most of the acid is not affected. This is why ponds in the Adirondacks became acidified (non-calcite rock in those areas), while Lake Champlain (abundant calcitic bedrock) did not.

The degree of acidification is the pH of the water, which is defined as the negative logarithm of the concentration of hydrogen ion (H^+), or

$$pH = -\log [H^+].$$

(This to a certain degree comes from the old definition of an acid as a proton donor. A hydrogen ion is little more than a proton, so think of it as the amount of free protons floating around).

A pH of 7 is considered neutral, while a pH less than 7 is considered acidic. For example, wine has a pH of about 3.5 and your stomach digestive fluids have a pH of about 1.9.

We should also be aware that increased acidity does not have to be constant, but instead can be episodic. High surface water discharge events (storms, snowmelts) can increase the pH of streams and ponds to dangerous levels for short times.

Effects of Acidity on Plants and Animals

As a first example of the effects of acid rain, we can examine a case which is not obvious–effects on non-aquatic, tree nesting birds. This study was carried out in the

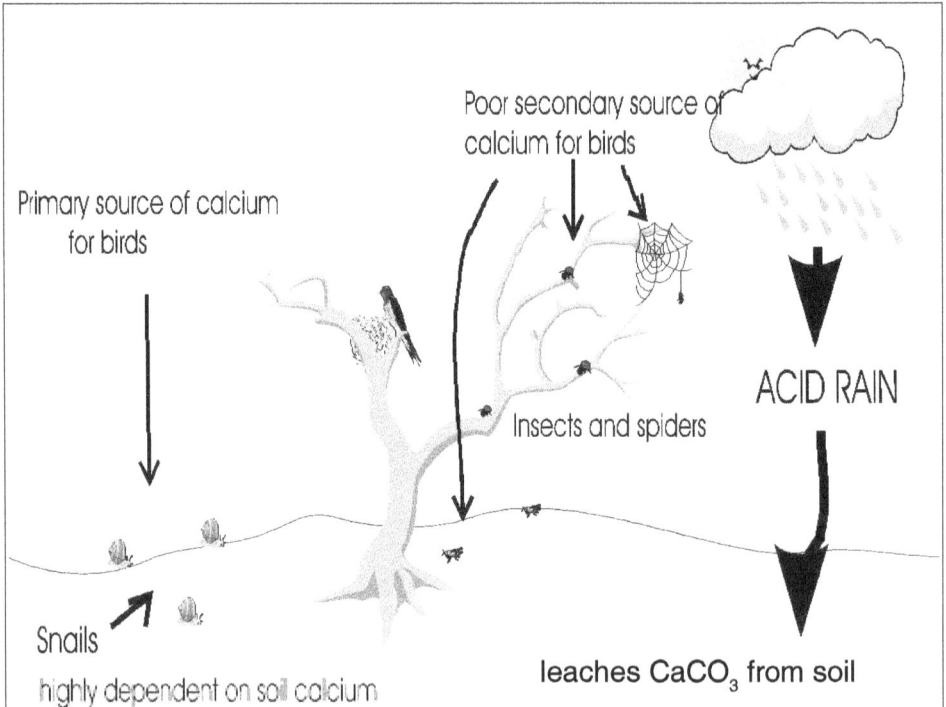

Figure 4

Netherlands. It was observed that the proportion of birds laying defective eggs rose from roughly 10 per cent in 1983-84 to 40 per cent by 1987-88. The defective eggs had thin and highly porous egg shells, which resulted in eggs failing to hatch because of shell breakage and desiccation. As a result, there was also a high proportion of empty nests and clutch desertion. It was also observed that these effects were limited to areas of acid rain.

Since the birds did not appear to be directly affected by the acidity, the food chain was examined (these birds are positioned at the upper part of the local food chain). The difference between areas of normal soil pH (buffered by high calcium content due to limestone and marble outcrops and bedrock) and those with acidic soil appeared to be the presence of snails. The snails depend on the soil as their calcium source as they secrete their shells. With much of the $CaCO_3$ leached out of the soil by the acid precipitation, the snails could not survive in the area. The birds did not, at first, appear to be affected, because they continued to eat spiders and insects which, while supplying a sufficiently nutritious diet for the birds, where a poor source of calcium.

To test the hypothesis that the lack of calcium was the cause of the bird's laying defective eggs, ecologists "salted" the area with chicken egg shell fragments. The birds began to eat the chicken egg shells, and those that did laid normal eggs.

In this case, acid precipitation had affects that passed on up the food chain.

Affects on Aquatic Systems

Mollusks–Snails and Clams

☆ These invertebrates are highly sensitive to acidification because of their shells which are either calcite or aragonite (both forms a $CaCO_3$) which they must take from the water.

☆ In Norway, no snails are found in lakes with a pH of less than 5.

☆ Of 20 species of fingernail clams, only 6 were found in lakes with pH of less than 5.

Arthropods

☆ Crustaceans are not found in water with a pH less than 5.

☆ Crayfish are also uncommon in water where the pH is less than 5.

☆ This is an important consideration because crayfish are an important food source for many species of fish.

☆ Many insects also become rare in waters with a pH less than 5.

Amphibians as you may know, many species of amphibians are declining. To what extent acid rain is contributing to this decline is not exactly known. However, one problem is that in places like northeastern North America amphibians breed in temporary pools which are fed by acidified spring meltwater. In general, eggs and juveniles are more sensitive to the affects of acidity.

Zooplankton in lakes- changes in diversity among zooplankton have been noted in studies carried out in lakes in Ontario, Canada. These studies found that in lakes where the pH was greater than 5 the zooplankton communities exhibited diversities of 9–16 species with 3–4 being dominant. In lakes where the pH was less than 5, diversity had dropped to 1–7 species, with only 1 or 2 dominants.

Periphytic algae- many acidified lakes exhibit a large increase in the abundance of periphytic algae (those that coat rocks, plants and other submerged objects). This increase has been attributed to the loss of heterotrophic activity in the lake (*i.e.*, the loss of both microbial and invertebrate herbivores in the lake).

Fish-as a result of acidification, fish communities have suffered significant changes in community composition attributed to high mortality, reproductive failure, reduced growth rate, skeletal deformities, and increased uptake of heavy metals.

Mortality- effects on embryos and juveniles:

Atlantic salmon fry have been observed to die when water with pH 5 was introduced into breeding pools.

- ☆ In fish embryos, death appears to be due to corrosion of epidermal cells by the acid. Acidity also interferes with respiration and osmoregulation. In all fish at a pH of 4 to 5 the normal ion and acid/base balance is disturbed. $Na+$ uptake is inhibited in low pH waters with low salinity. Small fish are especially affected in this way because due to their greater ratio of body and gill surface area to overall body weight, the detrimental ion flux proceeds faster.

- ☆ In all fish low pH water causes extensive gill damage. Gill laminae erode, gill filaments swell, and edemas develop between the outer gill lamellar cells and the remaining tissue.

- ☆ At pH <3 coagulation of mucus on gill surfaces clogs the gills, which leads to anoxia and subsequent death.

Reproductive Failure

Reproductive failure has been suggested as the main reason for fish extinction due to acidity. In Ontario, Canada it was observed that in acidified lakes female fish did not release ova during mating season. When examined, the fish were found to have abnormally low serum calcium levels which appears to have disrupted their normal reproductive physiology.

Growth

Growth may increase or decrease depending on resistance of a species to acidity. For resistant species, growth can increase due to the loss of competing non-resistant species. On the other hand, growth can decrease due to increase in metabolic rate caused by sublethal acid stress. In this case the organism's rate of oxygen consumption goes up because the excess CO_2 in the water increases the blood CO_2 level which decreases the oxygen carrying capacity of the hemoglobin.

Skeletal Deformity

This occurs in some fish as a response to the lowered blood pH caused by increase in CO_2 described above. Bones decalcify in response to a buildup of H_2CO_3 in the blood as the body attempts to maintain its normal serum osmotic concentration (*i.e.*, the body attempts to return to a normal blood pH level).

Almost everybody has heard about acid rain and knows that it is something bad. But what exactly is it? What are its effects on plants, animals, human beings, and what can be done to solve this problem?

The term acid rain does not convey the true nature of the problem and therefore scientists use the term "acid depositions". This is because the acid which has formed due to pollution may return to the earth as a solid or a gas and not just as rain. Depending upon the climatic conditions it could also come down as rain, fog, or snow, and in the wet form it is known as "acid precipitation".

Sources

Certain industries, as well as emissions from vehicles give rise to increase of sulphur dioxide and nitrogen oxides in the air. These emissions change into sulphates and nitrates under the influence of sunlight and moisture, and get converted into sulphuric acid and nitric acid, which come down as acid rain.

Coal generally contains between 2–3 per cent sulphur, and when it is burned, this sulphur is released into the atmosphere. Electric companies and other industries which burn coal cause a lot of emission of sulphur dioxide. Other industries which process raw ore containing sulphides in order to obtain copper, zinc, or nickel also cause an increase in sulphur dioxide levels in the atmosphere.

The major source of emissions of nitrogen oxides into the air, is from vehicles and other places where fossil fuels are burned. Forest fires, often caused by man, either deliberately or accidentally, are another source of pollution.

Naturally occurring phenomena like volcanic activity, lightning, or organic decay, also gives rise to an increase in atmospheric pollutants but not much can be done about these causes. However over 90 per cent of sulphur dioxide emissions and around 95 per cent of nitrogen oxides released into the air are from man made sources.

The problem of acid rain is not new. It was first noticed during the 17th century, when people observed the effects of industrialisation on plants and animals. As far back as in 1872, the Scottish chemist, Angus Robert Smith wrote a book "Air and Rain: The Beginnings of Chemical Climatology", in which he used the term "Acid Rain", and the name has stuck. The problem has become severe since the 1960s when fishermen noticed a sharp reduction in the quantity of fish in lakes of North America and Europe.

The havoc done by acid rain is not localised in the place where it is caused. The atmospheric emissions may travel for several days and over long distances depending upon wind and climatic conditions, before coming down as acid rain. The problem caused in an industrialised area may therefore result in acid rain in the surrounding forests or lakes, or even further away. It is believed that around 50 per cent of the acid rain that occurs in Canada is due to pollution caused in the United States of America, and the effect of polluting industries in England can be felt in Norway.

If there were no pollution, the rain would still be acidic. Natural rainfall has a pH of around 6.0. This is because of the effect of Carbon dioxide in the air which combines with water to form carbonic acid. The effect of this is however negligible, as it is neutralised in the soil by alkaline material like limestone. However the other emissions cause the pH of the rain water to drop below 5.5 and at this level it is considered to be acid rain. The soil cannot now neutralise the acidity of the rain water. In some places the acidification is so severe that the pH drops to around 4.0. Rare cases have been reported of acid rain having pH of around 2–2.5.

Effects of Acid Rain on Aquatic Life

The action of acid rain causes harmful elements like mercury and aluminium to be leached from the soil and rocks and it is then carried into the lakes where aquatic life may be affected. Warning signs have been posted at several lakes, telling about the dangers of eating fish which may have been poisoned by mercury. Just as the soil has a natural ability to neutralise the acidity of rain water, within a certain limit, so also lakes and other water bodies can to a certain extent nullify the effects of acid rain. However as the acidity increases, the natural mechanisms are no longer able to cope. As the water gets more acidic its pH goes down. As the pH reaches 5.5, plankton, certain insects and crustaceans begin to die. At a pH of around 5.0, the fish population begin to die. When the pH drops below 5.0, all the fish have died, and the bottom of the lake lies covered with undecayed material. Every year during the spring thaw, there is a sudden increase in the acidity of the lakes as frozen acid is suddenly deposited in them. This "Acid Shock" prevents the reproduction of aquatic species, or results in the deaths of the hatchlings.

Effects on Animals and Birds

All living organisms are interdependent on each other. If a lower life form is killed, other species that depended on it will also be affected. Every animal up the food chain will be affected. Animals and birds, like waterfowl or beavers, which depended on the water for food sources or as a habitat, also begin to die. Due to the effects of acid rain, animals which depended on plants for their food also begin to suffer. Tree dwelling birds and animals also begin to languish due to loss of habitat.

Effects on Human Beings

Mankind depends upon plants and animals for food. Due to acid rain the entire fish stocks in certain lakes have been wiped out. The economic livelihood of people who depended on fish and other aquatic life suffers as a result. Eating fish which may have been contaminated by mercury can cause serious health problems. In addition to loss of plant and animal life as food sources, acid rain gets into the food we eat, the water we drink, as well as the air we breathe. Due to this asthmatic people and children are directly affected. Urban drinking water supplies are generally treated to neutralise some of the effects of acid rain and therefore city dwellers may not directly suffer due to acidified drinking water. But out in the rural areas, those depending upon lakes, rivers, and wells will feel the effects of acid rain on their health. The acidic water moving through pipes causes harmful elements like lead and copper to be leached into the water. Aluminium which dissolves more easily in acid rain as compared to pure rainfall, has been linked to Alzheimer's disease. The treatment of urban water supplies may not include removal of elements like Aluminium, and so is a serious problem in cities too.

Other Effects

All living things, whether plants or animals, whether living on land or in the water or trees, are affected either directly or indirectly by acid rain. Even buildings, bridges and other structures are affected. In cities, paint from buildings have peeled off and colours of cars have faded due to the effects of acid rain. From the Taj Mahal in India to the Washington Monument great buildings all over the world have been affected by the acid rainfall which causes corrosion, fracturing, and discoloration in the structures. In Europe, structures like The Acropolis in Greece and Renaissance buildings in Italy, as well as several churches and cathedrals have suffered visible damage. In the Yucatan peninsula in Mexico, and in places in South America, ancient Mayan Pyramids are being destroyed by the acid rain. Temples, murals, and ancient inscriptions which had previously survived for centuries are now showing severe signs of corrosion. Even books, manuscripts, paintings, and sculpture are being affected in museums and libraries, where the ventilation system cannot eliminate the acid particles from the air which circulates in the building. In some parts of Poland, trains are required to run slowly, as the tracks are badly damaged due to corrosion caused by acid rainfall.

Solutions

The bottom line is that all things on earth are being affected by this problem and the good news is that something is being done to solve it. Pressure from the environmental groups, and public has increased as the effects of the havoc caused by acid rain become more apparent. Governments all over the world have drawn up plans to tackle this problem.

Lakes that have become highly acidic, can be treated by adding large quantities of alkaline substances like quicklime, in a process called liming. Although it has worked in several places, it has not been successful where the lake is very large, making this procedure economically unfeasible, or in other lakes where the flushing rate of the lake waters is too large resulting in the lake becoming acidic again.

The best approach seems to be in prevention. To this end environmental regulations have been enacted to limit the quantity of emissions released in the atmosphere. Several industries have added scrubbers to their smoke stacks to reduce the amount of sulphur dioxide dumped in the atmosphere. Specially designed catalytic converters are used to ensure that the gases coming out from exhaust pipes of automobiles, are rendered harmless. Several industries which use coal as fuel have begun to wash the coal before using it thereby reducing the amount of Sulphur present in it, and consequently the amount of emissions. Usage of coal with a low Sulphur content also reduces the problem.

We as individuals can take several steps to alleviate the effects of this problem. A reduction in use of vehicles will reduce the amount of emission caused by our vehicles. So do not use the car unless it is absolutely required. For going short distances, walk or try to use a bicycle. This will not only protect the environment but also improve your health. If the distance is greater, try using public transportation. If you must use your vehicle try forming a car pool and share your vehicle with someone else. Ensure that your vehicle is properly tuned, and fitted with a catalytic converter, to reduce the emissions.

Reduce use of electric power. Switch off lights, and other electrical appliances when not required. Do not leave your Televisions, VCRs, Microwave Ovens or Music Systems on Stand-by when not required. Switch them off.

Reducing power consumption will reduce the amount of coal burnt to produce electricity, and thus reduce the amount of pollution. This is true even if your electricity company does not use coal for producing electricity, but some other more environmentally friendly way. This is because the electricity you have saved can now be used elsewhere, thus benefiting nature.

Speak to others about this problem. Increasing awareness is one way of ensuring that things are done to solve this global problem. Find out what fuel is being used by your electricity company to produce electricity. If they use coal, ask what methods they use to contain, if not eliminate, the problem of sulphur emissions. Washing the coal used, or using coal having a low sulphur content, is costly and therefore some companies try to avoid this. If you have the option, switch to a utility that shows more concern for the environment.

Write to your representative in Government. Pressure from people can make Governments enact suitable legislation, to ensure that industries keep their emissions within limits. Join some group which works to protect the environment. When people get together and speak with one voice they are more likely to be heard.

Chapter 13

Global Environmental Issues

Some environmental issues have global significance and need to be addressed through international effort. This Subject Map outlines the four key issues of climate change, ozone depletion, biodiversity and sustainable development, outlining the problem, policy response and key sources and players in each case. Future research briefings will consider individual issues in greater depth.

Greenhouse Effect and Climate Change

A 'greenhouse layer' of gases that retains heat envelops the Earth. The 'greenhouse effect' is absolutely crucial in maintaining life on Earth since without this heat-retaining blanket we would be 33°C cooler. The main 'greenhouse gases' (GHGs) are water vapour and carbon dioxide (CO_2) although chlorofluorocarbons (CFCs), methane (CH_4), nitrous oxides (N_2O) and ozone (O_3) are also important.

The problem is that levels of these gases have risen significantly since preindustrial times or around 1750. This can be attributed largely to human activities, mostly fossil-fuel use and land-use change, and is exacerbated by the removal of traditional 'sinks' such as forests, fossil fuels and ocean plankton, that kept carbon fixed and 'locked up'.

The overwhelming consensus is that warming is occurring because by raising the levels of greenhouse gases, humans are enhancing ('forcing') the greenhouse effect. The Intergovernmental Panel on Climate Change (IPCC) is a massive international scientific collaboration of respected climate and other scientists, established in 1988 by the World Meteorological Organisation (WMO) and the United Nations Environment Programme (UNEP). In December 1995 the IPCC's Second Assessment confirmed that

"The balance of evidence … suggests that there is a discernible human influence on global climate".

The Report projected that global mean surface temperatures would increase by between 1 and 3.5°C by 2100, the fastest rate of change since the end of the last ice age. Global mean sea levels would rise by between 15 and 95 cm by 2100, flooding many low-lying coastal areas. Changes in rainfall patterns are predicted, increasing the threat of drought, floods or intense storms in many regions. According to the WMO the seven warmest years globally in the instrumental record occurred during the 1990s. Extremely heavy precipitation had devastating consequences in many parts of the world in 1999, while other areas were plagued by drought conditions. The recent heavy flooding in the UK has brought a timely reminder of the potential impacts of extreme weather conditions, together with more of an acceptance that human activities can impact on climatic regimes.

The UK Climate Impacts Programme (UKCIP) has examined likely effects of warming in the UK, and a research study let by the Executive to scope the implications of climate change in Scotland outlines UKCIP predictions; Scotland will become warmer, with relatively more warming in winter than summer. Rainfall is likely to increase, with autumn and winter seeing the biggest increases while spring amounts will drop and summer rainfall will remain similar to today. The intensity of rainfall events is likely to increase, leading to increased risk of flooding. The scenarios suggest a decrease in the number of gales overall, though there may be an increase in the frequency of very severe gales. The water balance is likely to remain favourable, which is important for the water industry and agriculture, while direct short-wave solar radiation is likely to reduce over the next century with increased cloud cover. Natural variability of the climate system is likely to modify the magnitude and patterns of these human-induced changes. Two possible 'climate surprises' are the collapse of the ocean circulation in the North Atlantic, with dramatic consequences for European climate, and the collapse of the West Antarctic ice sheet, which would increase the rate and magnitude of sea level rise. Neither is thought likely to occur in the near future.

The IPCC has recently updated its emissions scenarios. The main future drivers of change will be demographics, social and economic development, and technological change. Their interactions will be dynamic and complex. The scenarios will inform the IPCC's Third Assessment, due in May 2001. A draft of the Third Assessment reportedly says that average global temperatures could rise by twice as much as previously thought, possibly by 6°C above 1990 levels by 2100.

Policy Response

An important development prior to the formation, in 1998, of the IPCC, was the UN Framework Convention on Climate Change (UNFCCC), adopted in New York in 1992 and opened for signature at Rio or the 'Earth Summit' in June that year. The Convention sets an 'ultimate objective' of 'stabilising greenhouse gas concentrations in the atmosphere at a level that would prevent dangerous anthropogenic interference with the climate system'.

The Kyoto Protocol to the UNFCCC adopted in December 1997 set, for the first time, legally binding emission targets. The UK signed this in April 1998 but has not yet ratified. The Parties to the Protocol must, individually or jointly, reduce overall emissions of GHGs by at least 5 per cent from 1990 levels in the 'commitment period', which is 2008 to 2012.

Parties may 'burden share' to meet this target, so the EU will reduce emissions by 8 per cent from the 1990 baseline over 2008-12, while the USA agreed to a 7 per cent cut and Japan and Canada to 6 per cent. Some developing countries, in consequence, were allowed to increase their emissions compared to 1990 levels. The Department of the Environment, Transport and the Regions website gives details. The UK's target (which will become legally binding when we ratify) will be a 12.5 per cent cut towards the EU's overall 8 per cent target shared out among the 15 Member States. However, the UK has gone further and set a 'domestic goal' of a 20 per cent cut.

The Executive is responsible for environment policy, but many of the fiscal and energy policy measures needed to counter climate change are reserved. The Scottish Climate Change Programme Consultation issued in March 2000 agrees to work in partnership with the UK Government to achieve a 'domestic goal' of a 20 per cent reduction in CO_2 emissions by 2010 as set out in the draft UK Climate Change programme. The Executive's climate change web pages provide updates.

From 13-24 November 2000 the parties to the Convention will hold their sixth conference of the parties in The Hague, Netherlands. It is hoped the conference will speed the entry into force of the Kyoto Protocol by 2002. Key issues to be considered by the conference are: emissions trading systems, rules for obtaining credit for improving carbon 'sinks', monitoring of compliance with commitments, technology transfer, and the special concerns of developing countries that are particularly vulnerable to climate change or economic instruments.

Ozone Depletion

Ozone (O3) is a dangerous street level pollutant, contributing to photochemical smogs. It is also a minor greenhouse gas, contributing to climate change. But the main reason people are aware of its existence is through its beneficial effects in the stratospheric 'ozone layer' 5 to 30 miles up, which provides an important protection for life on Earth from the dangerous effects of the sun's radiation, by absorbing biologically damaging ultraviolet sunlight (UV-B).

The hole in the ozone layer was first noticed over the British Antarctic Survey station Halley, Antarctica. Today, up to 60 per cent of the total overhead amount of ozone is depleted during the Antarctic spring. In the Arctic Polar Regions a similar but smaller hole has appeared in 6 out of the last 9 years. The UNEP website provides information on how the holes form. NASA announced on 7 September 2000 the largest ever-Antarctic hole; three times the size of the USA. Increases in surface UV-B radiation have been observed in association with local decreases in stratospheric ozone, from both ground-based and satellite-borne instruments.

Ozone-depleting compounds are a group of chemicals called halocarbons that can contain the elements chlorine, fluorine, bromine, carbon, and hydrogen. As early as 1974 an article in *Nature* had shown that compounds being added to the Earth's atmosphere were destroying the ozone layer. One group of halocarbons called chlorofluorocarbons (CFCs) invented in 1928 found use in aerosols, foams, refrigeration, air conditioners, cleaning of electronic components, and as a solvent. Another group (halons) was used in fire extinguishers. Once released, halocarbons are long-lived and stable chemicals that rise up and persist in the stratosphere for many years, where they break down ozone.

Policy Response

The UN started to address the problem in the 1970s, resulting in the 1987 Montreal Protocol on Substances that Deplete the Ozone Layer, made under the 1985 UN Vienna Convention for the Protection of the Ozone Layer. The Protocol aims to reduce and eventually eliminate the emissions of man-made ozone depleting substances, by stopping their production and use, and has been modified or strengthened five times so far by amendments. As more 'culprit' depleting substances are identified, the scope of the Protocol has been expanded.

An EC Regulation directly applicable in UK law implements the Protocol in the UK. EC Regulation 2037/2000, effective from 1 October 2000 implements the latest rules, such as an immediate ban on the sale and use of most CFCs and some other chemicals, although exemptions apply. The use of hydrochlorofluorocarbons (HCFCs) in most new refrigeration and air conditioning equipment is prohibited from 1 January 2001, and certain fire fighting systems that use other halons are prohibited from 31 December 2003.

This is, to some extent, a success story. The UN Secretary General said in September 2000 that without the Protocol, the levels of ozone-damaging substances would have been five times higher than they are today, but developing countries are yet to phase out CFC emissions to meet the 2010 deadline imposed by the Montreal Protocol. There are also reports of a black market in CFCs.

According to a recent UNEP news release, scientists predict that the ozone layer will fully recover some time in the 21st century – 'but only if the Protocol continues to be vigorously enforced'.

Biodiversity

Biodiversity means the whole variety of life on Earth. The 1992 UN Biodiversity Convention defines biodiversity as 'the variability among living organisms from all sources including terrestrial, marine and other aquatic ecosystems and the ecological complexes of which they are part; this includes diversity within species, between species and of ecosystems'. Thus genetic diversity is included, which is increasingly important given developments in biotechnology and farming.

No one knows how many species exist, because relatively few have been found and even fewer named. (This naming – or taxonomy – is important because it gives us some idea of a species' place in evolution and interdependencies.) Estimates range from 7 million plants and animals to 14 or 30 million. UNEP's Global Biodiversity Assessment (by a multinational collection of thousands of scientists) estimated that species extinction has been occurring at 50 to 100 times the average natural rate since 1600. Many remaining species are concentrated into remaining 'pockets' of habitat; 30-50 per cent of plant, amphibian, reptile, mammal and bird species occur in 25 'hotspots' around the globe, taking up around 2 per cent of land surface. When these pockets go the extinction rate is expected to rise by thousands of times the natural rate. Over 31,000 plant and animal species are now threatened with extinction and over 10 per cent of plants and 10 per cent of birds are expected to become extinct in the wild in the medium-term.

The five major causes of biodiversity loss are

1. Fragmentation, degradation or outright loss of habitats;
2. Over-exploitation of biological resources;
3. Pollution;
4. Introduction of non-native species (rabbits to Australia, for instance);
5. Climate change.

The habitats causing most concern are forests (not just tropical but temperate forest), marine and coastal areas, and agricultural and inland water ecosystems. The World Resources Institute estimates that from 1960 to 1990, one fifth of all natural tropical rain forest cover was lost. Up to 10 per cent of the world's coral reefs have already been degraded beyond recovery. 50 per cent of the world's coastal mangroves have been cleared. The UN Food and Agriculture Organisation's (FAO) State of the World's Forests says that from 1980 to 1995 developing nations lost a net 200 million hectares of forest. Over 30 per cent of animal breeds are threatened. At the same time, clearing land for agriculture is a major source of biodiversity loss. From 1700 to 1980 the amount of cultivated land grew from 5 per cent to 35 per cent of total land area. The UK has lost over 100 species this century and many more have declined in number, range, or both. The major causes of these losses are inadequate or inappropriate land management, air or water borne pollution and a poor definition of who is responsible for furthering sustainable development.

Policy Response

The UN 1992 Convention on Biological Diversity was one of the main outcomes of the Rio or Earth Summit, along with the Climate Change Convention. Its aims are 'the conservation of biological diversity, the sustainable use of its components and the fair and equitable sharing of the benefits arising out of the utilisation of genetic resources.' This recognises then that biodiversity is a resource, as well as a responsibility, and encourages international co-operation, such as the transfer of environmentally sound technologies, to preserve it.

The UK signed the Convention in June 1992 and ratified in June 1994. As required by the Convention, a UK Biodiversity Action Plan was published in 1994, setting out a broad strategy and approach. A UK Biodiversity Group took over from an initial steering group, and the Scottish Biodiversity Group (SBG) was established in May 1996. Action for Scotland's Biodiversity describes the unique Scottish inheritance of species and habitats and outlines ways of protecting this.

The number of costed biodiversity 'action plans' has risen from a first set of 130 published in 1995 to 436 by the end of 1999, covering all the UK's identified 'priority' species and habitats. The final action plans were launched in October 1999. 226 species action plans and 41 habitat action plans are relevant to Scotland, and plans for species or habitats endemic and unique to Scotland within the UK include native pinewoods, capercaillie, the Scottish crossbill and mountain scurvy-grass. The action plans set down detailed conservation targets, the actions necessary to meet them, and the government organisations responsible for taking this action. The SBG has issued a manual to guide local authorities and others involved in preparing and implementing Local Biodiversity Action Plans.

Sustainable Development and Agenda

The World Resources Institute, which has just published *World Resources 2000-01* says that world population, now at 6 billion, is projected to grow to 8-12 billion in 2050. Nearly all of this growth is expected in the developing world. Some 800 million people remain undernourished. Global energy use has risen by almost 70 per cent since 1971 and is projected to increase at over 2 per cent each year for the next 15 years. Global water consumption may become one of the most contentious resource issues of the 21st Century; a third of the world's population already lives in countries experiencing moderate to high water stress.

The Rio Declaration produced after the Earth Summit in Rio de Janeiro in June 1992 has a common theme of sustainable development running through it. The 1987 Brundtland Commission first defined sustainable development as 'development that meets the needs of the present without compromising the ability of future generations to meet their own needs'.

Sustainable development is increasingly shifting from a purely 'environmental' to a more social issue. Before Rio, the UK's environmental strategy, set out in the *This Common Inheritance* series of white papers, included a definition of sustainable development that today we would see as very confined to 'environmental' issues;

> Sustainable development means living on the earth's income rather than eroding its capital. It means keeping the consumption of renewable natural resources within the limits of their replenishment. It means handing down to successive generations not only man-made wealth, but also natural wealth, such as clean and adequate water supplies, good arable land, a wealth of wildlife, and ample forests.

Some critics say that sustainability has been seen as environmentalism in another guise and not adopted by developing countries, because Western nations have been paying too little attention to the chapters of Agenda 21 on poverty, shelter and livelihoods. A right to develop is seen as crucial by non-industrialised countries, with sustainable development not being about stopping growth. Indeed the Earth Summit's full name was the UN Conference on Environment *and Development* (UNCED). 'Poverty and social exclusion' is now one of the UK's headline measures of sustainable development and economic growth is explicitly included in the UK's latest sustainable development strategy;

Although the idea is simple, the task is substantial. It means meeting four objectives at the same time, in the UK and the world as a whole:

☆ Social progress which recognises the needs of everyone;

☆ Effective protection of the environment;

☆ Prudent use of natural resources; and

☆ Maintenance of high and stable levels of economic growth and employment.

At the same time 'environmentalism' is still key – from loss of biodiversity and ecosystems to ozone depletion, fossil fuel use, over use of nitrogen fertilisers, invasion of non-native species, acid rain, deforestation and forest fires, global warming and water stress; the issue is complex and all-encompassing.

Policy Response

A forty chapter 'sustainable action plan' called Agenda 21 was one of the main products of Rio, but while signed by 153 countries it has no legal status (unlike the Biodiversity and Climate Change conventions). The UK produced its first Sustainable Development Strategy as a follow-up to Agenda 21 in 1994. This promised a set of 'sustainable development indicators', to show whether the UK was moving towards a more sustainable way of life. A first set of Indicators of Sustainable Development for the UK was produced in 1996.

Two bodies were established; a Government Panel to advise Ministers and Round Table with members from sectors of society and business. A Scottish Advisory Group on Sustainable Development was established in 1994. This was wound up in 1999, ending with ten 'action points' for the new parliament, including the recommendation that a "Sustainable Development Commission" be established. The new Labour Government's sustainable development strategy for the UK 'A better quality of life', promised to publish progress against headline sustainable development indicators annually, to be measured by a new Sustainable Development Commission combining the Round Table and Panel on Sustainable Development.

Quality of Life Counts outlines the current 15 headline indicators, some with specific targets, but the devolved administrations were left to set their own indicators. In January 2000, the Ministerial Group on Sustainable Scotland was established, and now has members including the Ministers for Transport, Environment, Finance and Social Justice, as well as external representatives from Friends of the Earth Scotland and Shell Expro. In the February 2000 Parliament debate on sustainable development, then Minister for Transport and the Environment Sarah Boyack announced that Scotland would have its own set of headline indicators. The Minister also outlined the main priorities of the Executive's strategy, which are Waste, Energy and Travel (WET).

The new UK Sustainable Development Commission (not to be confused with the UN Commission for Sustainable Development) was launched in October 2000 with Jonathon Porritt as its chair, and welcomed by Sarah Boyack.

At a local level, the Prime Minister set a target that all local authorities should prepare local sustainable development or Local Agenda 21 (LA21) strategies by the year 2000. COSLA's Scottish Local Agenda Survey of March 2000 found that of 31 Councils that replied, 26 thought they would adopt a strategy by December 2000, two had adopted a strategy and two intended to. One council would have no strategy in place by the deadline and one failed to respond to the survey. But the detail was lacking; 60 per cent of councils could not provide a full time officer to support their LA21, and 75 per cent could not show the detailed steps they would take to adopt a Local Agenda 21 strategy by the end of 2000.

Issues in Environment Policy Consumption versus Preservation of Environmental Resources

Sustainable development goes beyond a static framework of correcting environmental externalities, deficient markets and inadequate property rights. It is very much concerned with the availability of environmental and natural resources in the future. The problem may be particularly acute in developing countries that have high discount rates due to the scarcity of capital and prevalence of poverty (which indicates a strong preference for present over future consumption). Therefore, use of an appropriate discount rate/shadow prices is central to maintaining an optimal balance between consumption and preservation of natural resources.

Valuation of Environmental Damages

Attempts to establish optimal pollution abatement levels or to set more general environmental quality objectives confront several problems in determining the monetary value of environmental damage. One difficulty in setting optimal abatement levels is that an economic activity must be functionally related to measurable environmental changes and the scientific basis for establishing the linkages may not exist. Secondly, only a use value of a good is normally accepted as contributing to welfare. Environmental goods have non-use value and option value (willing to pay for preserving the option to use the goods in future), also. Thirdly, many environmental services do not pass through markets and, therefore, no market prices exist to indicate value. Determining the value also depends on the type of environmental damage under consideration (damage to productive resources, to eco-system maintenance services or human health). Among the economic valuation techniques adopted for such valuation of environmental damage are the contingent valuation method, travel cost method, hedonic pricing, surrogate markets method etc. The importance of valuation of environmental damage lies in demonstrating the existence of substantial economic benefits from increased resource protection. An appropriate emphasis, therefore, needs to be placed on valuation of such environmental damages and its suitable internalisation/incorporation in the development projects.

Natural Resource Accounting

Accounting schemes that more accurately measure the environmental costs of economic activity at all levels could help increase the efficiency of natural resource use and reduce related environmental impacts. The need for Natural Resource

Accounting (NRA) and their integration with the system of national accounts have been emphasised in various policy documents of the Ministry of Environment and Forests. The Ministry in 1993, had developed a framework for preparing such integrated accounts for India. It has further taken up programme for development of prototype accounts for air, water, biodiversity and common property resources for selected hotspots in the country with a view to firm up the methodologies and techniques in their preparation.

Use of Economic Instruments/Price Mechanism

While regulatory measures remain essential, new approaches for considering market choices in the protection of environment are being increasingly adopted by both developed and the developing countries. The aim is to give industries and consumers clear signals about the cost of using environmental and natural resources. If the environmental costs of economic activity are more adequately reflected in the prices paid for goods and services, then companies and ultimately the consumers would be guided to adjust their market behavior so as to reduce pollution and waste. The expectation is that price signals will influence behavior to avoid excessive use of natural resources/pollution build up. Hence, a greater use of such market based approaches as a complement to the present regulatory system is desirable. Use of such incentives would facilitate operationalising the standard approach of Polluter Pays Principle.

Removing Subsidies that Encourage Unsustainable Use

Production of natural resource related commodities is often subsidised. For example, Power tariff charged is only a fraction of costs of generating power. Such subsidies encourage wasteful use of resources, often creating or exacerbating environmental problems. Salination from overirrigation is a major cause of land degradation in the country. Imbalance in the use of fertilizer nutrients (NPK), especially owing to underpricing such as, urea affects soil productivity. Low price of kerosene has encouraged its adulteration especially of diesel and motor spirit contributing to air pollution. Similarly, the water and sanitation problem may be largely due to low user charges, starving the implementing agencies of funds to maintain and operate these facilities. Eliminating or reducing such subsidies would confer both economic and environmental advantages. However, in certain cases, elimination of such subsidies may require weeding out of inefficient and marginal enterprises, thus raising unemployment. A careful targeting of subsidies where these are most needed is, therefore, called for.

Extension of Property Rights

In some cases, the problem of environmental externality derives not from lack of appropriate user charges but from the lack of or ill defined property rights over resources. The extensive degradation of village commons or overexploitation of lakes/ oceans for fisheries are examples of such externalities. In such cases it may be more efficient for the government to assign or clarify property rights and allow private agents to handle problems of environmental quality through negotiations among the affected parties.

Trade and Environment

The concern with environmental implications of trade involves both the domestic implications of policy reforms as well as the global environmental dimension of international trade agreements. Domestic implications of trade policy would suggest growth of less polluting industries and shift to higher quality and environment friendly products/processes for traditional polluting industries (like leather and textiles) through adoption of cleaner technologies, so that greater efficiency and productivity associated with liberalised trade policy may also reduce pollution intensity. As regards global environmental dimension of international trade, the debate has revolved around the issue of whether free trade is beneficial to global and national environmental conditions and whether it should be used to influence national and international environmental standards and agreements. A number of countries have adopted trade related environmental measures with implications for product/process standards, packaging and labelling of export merchandise. These measures have the impact of non-tariff

Green Accounting

☆ National Accounts have been providing the most widely used indicators for the assessment of economic performance, trends in economic growth and the economic counterpart of social welfare. However, the new emphasis on sustainable development draws attention to the need for a broader assessment of growth and welfare by modified national accounts. In assessing cost and capital, national accounts do not consider scarcities of natural resources which threaten the sustained productivity of the economy and the degradation of environmental quality and consequential effects on human health and welfare. In addition, some expenditure for maintaining environmental quality are accounted as increases in national income and product. This is despite the fact that such outlays could be considered a maintenance cost of the society, rather than social progress. Thus, the conventional accounts are likely to send wrong signals and may result in policy decisions which are non-sustainable for the country. Green accounting on the other hand is, focused on addressing such deficiencies in conventional accounts with respect to the environment.

☆ Integrated environment and economic (green) accounting, therefore, attempts at accounting for both socioeconomic performance and its environmental effects and integrating environmental concerns into mainstream economic planning and policies. Such accounting imply allocating environmental costs (and benefits) to those activities and sectors that have caused them, in other words accounting for accountability, is a pre requisite for national management of both the environment and the country. Given the experimental nature of some of the proposed methodologies, particularly those on monetary valuation of non-marketed asset and externalities, such environmental accounting would require numerous and controversial estimates and valuation. Therefore, rather than modify the 'core' system of SNA, their incorporation through a system of "satellite accounts" have been suggested.

☆ Such integrated accounts can be useful in assessing the sustainability of economic growth and also the structural distortion of the economy by environmentally unsound production and consumption patterns. However, lack of international consensus on how to incorporate environmental assets and costs (and benefits) of their use in national accounts and existence of low statistical capacities for measuring natural resources depletion and environmental quality changes have resulted in a slow progress in development of green accounts.

☆ Nevertheless, the idea of placing statistical coverage of environmental concerns in a national accounts framework commands widespread support. Already, several attempts have been made at experimenting with satellite accounts–notably in Costa Rica, Mexico, the Netherlands, Norway and Papua New Guinea, among others. Indicative estimates suggest that conventionally measured GDP may exceed GDP adjusted for natural resources depletion and environmental degradation by between 1.5 per cent to 10 per cent.

Environmental Taxes

☆ As against the Command and Control approach to management of the Environment, the Economic or Market Based Instruments (MBIs) approach sends economic signals to the polluters to modify their behaviour. The approach normally involves financial transfers between polluters and the community and affects relative prices. But the polluters have freedom to respond and adjust, in the manner they want. They can thus choose the least cost option to meet the requirements. Hence, it is considered to be an efficient approach compared to the approach based on standards and regulations. The MBIs, therefore, have the benefit of being flexible and cost effective providing incentives for dynamic efficiency and resource transfer.

☆ Economic instruments used for environmental tax include pollution charges (emission/affluent tax/pollution taxes), marketable permits, deposit refund system, input taxes/product charges, differential tax rates and user administrative charges and subsidies for pollution abatement. These can be both price based and quality based instruments.

☆ MBIs have been applied in both developed as well as developing countries. In general, price based MBIs have been more widely used than those which are quantity based. Within price based MBIs, indirect instruments such as input-output taxes, differential tax rates and user fees have found extensive application in developed countries. By contrast, developing countries have made greater use of subsidies including those for end-of-pipe treatment equipment.

☆ The main MBIs used in India are subsidies for pollution abatement equipment for air and water resources. This provides rebates on duties for various pollution control equipment, monitoring instruments and abatement machinery for air/water pollution and promotion of unleaded fuel/fuel efficient automobile subsidy on automobile pollution kits/ converters etc. Accelerated depreciation for pollution control machinery is also provided. Among user charges/administration charges, consent fee is charged from industries under the Water Act and the Air Act. A water cess based on the consumption of water and type of industry (polluting) is also levied on selected industries and urban municipalities to conserve consumption and control pollution of water barriers and need to be tackled through improved product/process quality. Removing such trade barriers would increase employment, promote economic diversification in developing countries, and facilitate a greater effort towards improvement of the environment.

Development that can Reduce Poverty

Efforts should be made to encourage economic growth and provide employment to the poor. The growth strategy should also accord weight to the direct intervention programmes for poverty alleviation, stabilisation of population growth rate and greater employment opportunities for both the rural and urban poor. Further, it should target human capital investment in healthcare and education, provision of basic civic amenities like drinking water and sanitation, legal reforms to extend land tenure and other rights to the rural poor and enhancing the status of women while increasing their participation in development.

Peoples' Participation–Green Movement

Public participation is an essential ingredient of environment management along with other components like regulation and use of economic instruments. Public participation is facilitated by various environmental education and awareness programmes. The MOEF implements formal as well as informal education programmes and conducts various environmental awareness programmes. The usefulness of public participation is highlighted by the success of programmes like the Joint Forestry Management. A common approach followed by many countries, including India towards greater public participation is the scheme of Eco-labeling, which helps consumers to identify products that are environment friendly. Public hearing of major projects coming for environmental clearance to the MOEF has also been made mandatory. Other approaches in practice include the "consumer cooperatives" in Japan which popularise green products which are recyclable, biodegradable, rechargeable, ozone friendly and unleaded. Indonesian experiment with rating and public disclosure of the environmental performance of industries/ factories has also vindicated the effectiveness of public participation/opinion in management of the environment.

Participation in Global Dimensions of Environment

The country is signatory to various international Conventions having a direct bearing on environmental protection and conservation. Such Conventions play an important role in harmonising approaches and ensuring joint action on various global environmental issues. They also facilitate flow of larger funds and greater access to new technologies, contribute to protection of natural resource commodities and promote more sustainable production methods. Environmental protection and conservation of natural resources emerged as the key priorities in the wake of the Stockholm Conference on Human environment in 1972. The country is also a party to the Rio Declaration on Environment and Development and the Agenda 21, the operational programme for sustainable development

The United Nations Framework Convention on Climate Change (UNFCCC) aims at stabilization of greenhouse gas concentrations in the atmosphere at levels that would prevent dangerous anthropogenic interference with the climate system. It enjoins upon the parties to implement commitments contained in the various provisions of the Convention. As per the existing commitments, India is not required to adopt any reductions and limitation of greenhouse gas emissions. However, the third Conference of Parties to the Convention, held in Kyoto, Japan in December 1997, considered an alternate Protocol for strengthening of commitments of the parties to achieve the objectives of the Convention. The Kyoto Conference witnessed deliberate efforts of the developed countries to introduce new commitments for developing countries on the one hand, and to secure flexibility to implement their commitment through new mechanisms (such as emission trading, bank borrowing and joint implementation), on the other. The final agreement calls upon for an average cut in greenhouse gases emission of 5.2 per cent below 1990 levels to be achieved between 2008 and 2012 (European Union 8 per cent reduction, Japan 6 per cent reduction and U.S.A 7 per cent reduction). The provision of joint implementation and emission trading, however, would be confined to developed countries only. A clean development mechanism has also been defined for use by both developing and developed countries for sustainable development and implementation of commitments respectively.

The Convention on Biological Diversity attempted at conservation of biodiversity, sustainable use of its components and fair and equitable sharing of the benefits arising out of utilisation of genetic resources. Several steps, like preparation of a National Action Plan, framing a legislation on biodiversity, intellectual property rights on patenting of micro-organisms etc., have been initiated to meet our commitments under the Convention, as also to restructure administrative and policy regime in tune with the Articles of the Convention. Similarly, by acceding to the Convention to Combat Desertification, the country has reiterated its commitment to continue to accord priority to all actions to prevent land degradation and to improve the productivity of land on which the poor depend for their subsistence in such areas.

With a view to strengthen the global efforts to protect the ozone layer, India acceded to the Montreal Protocol in 1992. Draft rules on Ozone Depleting Substances (ODS) phaseout under the Environment (Protection) Act, 1986 have been prepared after extensive consultations with industry, NGOs and concerned Government Departments. The policy to issue licences for import of ODS has since been formulated. Duty exemptions are granted for new investment with non ODS technologies.

Instructions have been issued to all commercial banks prohibiting finance/refinance of new investments with ODS technologies.

Financing Sustainable Development

The share of environment and forests sector in the Eighth Five Year Plan public sector outlay amounted to 1.1 per cent, as compared to other social sector shares of 1.2 per cent for urban development, 1.3 per cent for housing, 1.4 per cent for family welfare, 1.6 per cent for medical and public health and 3.9 per cent for education. About 75 per cent of this approved outlay of environment and forests was in the state sector with bulk of the investment being in the forestry sector. The approved plan outlays in the Environment and Forests sector in the Seventh and Eighth Plans.

The United Nations Conference on Environment and Development (UNCED) secretariat has estimated that $ 600 billion would be required over the period 1993-2000 to implement Agenda 21 in developing countries. About two thirds of these resources are expected to come from the developing countries and the remaining as concessional aid from developed countries. The developed countries commitment to give 0.7 per cent of their GNP as concessional aid to developing countries for their sustainable development has not, however, materialised so far. Higher priority in resource allocation for this sector, therefore, need to be emphasised in order to implement its mandate on sustainable development programmes. Besides, greater involvement of the private sector and higher external aid would be needed to garner higher investment in this sector. For example, out of the total forest cover of 63.3 million hectare in the country, about 25 million hectare *i.e.*, 40 per cent is degraded. The current availability of funds is resulting in afforestation of around 1.2 million hectare only per year. Thus, at the present rate, it will take more than 20 years to afforest all the degraded forest land.

Keeping in view the resource constraints, the Ministry of Environment and Forests have identified in its Environment Action Programme priority areas covering conservation of and sustainable utilisation of Biodiversity in selected eco-systems; afforestation, wasteland development, conservation of soil and moistures including measures to lessen/eliminate water pollution; control industrial and related pollution with accent on reducing and/or management of wastes particularly those of hazardous nature ; improve access to cleaner technologies; tackling of urban environmental issues; and strengthening the scientific understanding of environmental issues as well as structures at different levels and orientation and creation of environmental awareness and an alternative energy plan with emphasis on cleaner sources of fuel.

Global Warming

The Earth's atmosphere has been maintained at a temperature that is comfortable for human, animal and plant life because of the balanced properties of carbon dioxide and other so-called "greenhouse gases" in the atmosphere. Increased human activity in recent years has released large additional amounts of these greenhouse gases into the atmosphere, however, exacerbating the global-warming effect. Of particular significance, the impact of carbon dioxide released through human activity since the

Industrial Revolution accounts for about 64 per cent of all factors that contribute to observed possible global warming worldwide.

According to the assessment report of the Intergovernmental Panel on Climate Change (IPCC) published in 2001, the average air temperature at the Earth's surface has been increasing since 1861. In the twentieth century, it increased by 0.6°C (± 0.2°C), the IPCC report predicted that the surface air temperature will rise an additional 1.4–5.8°C between 1990 and 2100. Many scientists think that this rapid increase of temperature will cause (1) sea levels to rise, (2) a widening of the economic disparities between developed and developing countries, (3) food crises as a result of decreasing agricultural production, (4) the disappearance of many wild species of animals and plants, and (5) negative impacts on human health due to an increase in the incidence of infectious disease.

Ozone Layer Depletion

Most of the Earth's ozone is produced and concentrated in the stratosphere, which extends from about 10 km to 50 km in altitude. It has been clearly proven that the stratospheric ozone layer has been depleted by ozone-depleting substances such as chlorofluorocarbons (CFCs), hydro chlorofluorocarbons (HCFCs), and methyl bromide. A depleted ozone layer means that harmful ultraviolet rays can penetrate more intensely to the Earth's surface. There are fears about their potential damage to human health by promoting skin cancer, cataracts, and so on, but also about harm to the growth of plants and plankton.

Because these ozone-depleting substances are chemically stable, they reach the stratosphere without being decomposed. In the stratosphere, they are decomposed by capturing ultraviolet radiation from the sun and releasing chlorine and bromine, which themselves act as catalysts for additional chain reactions that decompose the ozone.

Marine Pollution

It is not clear how the extent to which the vast expanse of the global oceans are polluted, because studies have mostly been conducted in the sea areas around developed countries and along highly populated coasts.

However, occurrences of "red tide" and other forms of pollution in enclosed seas, including the North Sea, the Baltic Sea and the Mediterranean Sea are increasing as a result of the emission of pollutants including hazardous substances and heavy metals. Serious marine pollution is also occurring as a result of the increase in the number of large tankers and offshore oilfield developments. The marine environment is seriously affected by large spills from oil tankers, because the effects are long-lasting and spread over a wide area. Protection of the marine environment is an important issue.

Transboundary Movements of Hazardous Waste

Inappropriate disposal and illegal dumping of hazardous waste exported from developed countries to developing countries caused environmental pollution in the

1970s and 1980s. There were many cases in which hazardous waste transporters were refused permission to land and forced to drift in the ocean with no destination. Hazardous waste was exported to developing countries that had relatively lax regulations and lower disposal costs. Issues concerning trans-boundary movements of hazardous waste need to be addressed globally, not only by developed countries but also by developing countries.

Deforestation

Forests are natural resources that provide various values. They provide habitats for wildlife and have environmental functions such as conservation of soil, enhancing the water holding capacity of the land, and absorption and sequestration of carbon dioxide. Furthermore, forests are a source of timber, firewood and charcoal, which are essential for human activities, as well as a resource to supply non-wood products such as the raw materials for pharmaceuticals.

Extensive exploitation reduced global forest cover between 1990 and 2000 by an estimated 94 million hectares, or about 2.4 per cent of total forest area (United Nations Food and Agriculture Organization, "State of the World's Forests 2001"). Natural forests of the tropical regions were estimated to have decreased by an average of 14 million hectares per year between 1990 and 2000, equal to an area about two-thirds the size of Honshu (the main island of Japan). About half of all wild species live in tropical forests, which could be considered as a genetic treasure trove, but there is grave concern that many species are in danger of extinction as a result of large-scale losses of tropical forest.

The reasons of tropical deforestation differ from region to region. Most forest clearing is done for agricultural purposes: traditional slash and burn agriculture, clearfelling of trees for sale as timber or pulp, grazing, plantation development and cash cropping. The general picture is complex, however, because of a mix of social and economic factors, such as population increase, poverty, and landuse systems.

Deforestation has recently been a concern in the Russian Far East and other middle-latitude areas of East Asia.

Some scientists suggest that the large amounts of carbon dioxide released as a result of deforestation accelerate global warming.

Loss of Biodiversity

The total number of species worldwide is estimated at about 30 million. Scientists have identified about 6,000 species of mammals, 9,000 species of birds, and 0.27 million species of vascular plants (pteridophyta and spermatophyte). Insect species account for most of the rest, although the most are still unknown.

Meanwhile, the pace of extinction of species caused by human activities has been accelerating. According to the Red List of Threatened Species compiled by the International Union for the Conservation of Nature and Natural Resources (IUCN), revised in 2000, there are currently 5,435 animal species and 5,611 plant species in danger of extinction. This situation is also evident in Japan, the home of about 90,000 species: 20 per cent of the approximately 240 mammal species, 13 per cent of the approximately 700 bird species, and 19 per cent of the approximately 8,800 vascular plants are at risk of disappearance.

Desertification

Although the term "desertification" is often only used to refer to the drying of the land, desertification phenomena also include soil erosion, salinization, and the loss of natural vegetation.

According to a 1991 study by the United Nations Environment Programme (UNEP), desertification affects about 3.6 billion hectares, representing 70 per cent of all dry land, and affects about 0.9 billion people – about one-sixth of the world's population.

Desertification can be caused by natural forces, such as drought, as well as by human activity, such as overgrazing and cultivation, excessive removal of trees for firewood and charcoal, and improper irrigation (which can cause the accumulation of salt in croplands). Solutions for the desertification problem are hampered in developing countries by many social and economic factors, such as poverty, population growth, foreign debt, and poor trade conditions.

Chapter 14
Indian Environmental Challenges

The key environmental challenges that the country faces relate to the nexus of environmental degradation with poverty in its many dimensions, and economic growth. These challenges are intrinsically connected with the state of environmental resources, such as land, water, air and their flora and fauna. The proximate drivers of environmental degradation are population growth, technology and consumption choices, and poverty, leading to changes in relations between people and ecosystems, and development activities such as intensive agriculture, polluting industry, and unplanned urbanisation. However, these factors give rise to environmental degradation only through deeper causal linkages, in particular institutional failures, resulting in lack of clarity or enforcement of rights of access and use of environmental resources, policies which provide disincentives for environmental conservation (and which may have origins in the fiscal regime), market failures, (which may be linked to shortcomings in the regulatory regimes), and governance constraints.

Environmental degradation is a major causal factor in enhancing and perpetuating poverty, particularly among the rural poor, when such degradation impacts soil fertility, quantity and quality of freshwater, air quality, forests, and fisheries. The dependence of the rural poor, in particular, tribal societies on their natural resources, especially biodiversity, is self-evident. The poor are particularly vulnerable to loss of resilience in ecosystems. Large reductions in resilience may mean that the ecosystems, on which livelihoods are based, break down, causing distress. The loss of the environmental resource base can result in certain groups of people being made destitute, even if overall, the economy shows strong growth. Further, urban environmental degradation, through lack of (or inappropriate) waste treatment and sanitation, industry and transport related pollution, adversely impacts air, water, and soil quality, and differentially impacts the health of the urban poor. This, in turn, affects their capability to seek and retain employment, attend school, and enhances gender inequalities, all of which perpetuate poverty.

Poverty itself can accentuate environmental degradation, given that institutional failures persist. For the poor, several environmental resources are complementary in production and consumption to other commodities (*e.g.*, water in relation to agricultural production, fuel-wood in relation to consumption of food), while a number of environmental resources are a source of income or food (*e.g.*, fisheries, non-timber forest produce). This is frequently a source of cumulative causation, where poverty, gender inequalities, and environmental degradation mutually reinforce each other. Poverty and environmental degradation are also reinforced by and linked to population growth, which in turn, depends on a complex interaction of diverse causal factors and stages of development.

Economic growth, in its turn, bears a dichotomous relationship to environmental degradation. On the one hand, growth may result in "excessive" environmental degradation through use of natural resources and generation of pollution aggravated by institutional failures. If impacts on the environmental resource base are neglected, an incorrect picture is obtained from conventional monetary estimates of national income. On the other hand, economic growth permits improvement in environmental quality by making available the necessary resources for environmental investments and generating societal pressures for improved environmental behaviour and institutional and policy change.

It is increasingly evident that poor environmental quality has adversely affected human health. Environmental factors are estimated as being responsible in some cases for nearly 20 per cent of the burden of disease in India and a number of environment-health factors are closely linked with dimensions of poverty (*e.g.* malnutrition, lack of access to clean energy and water). It has been established that interventions targeted at environmental management – *e.g.* reducing indoor air pollution, protecting sources of safe drinking water, sanitation measures, improved public health governance – offer tremendous opportunities in reducing the incidence of a number of critical health problems. It is also evident that these environmental protection measures would be difficult to accomplish without extensive awareness raising and education.

Institutional failures, referring to unclear or insufficiently enforced rights of access to and use of environmental resources, result in environmental degradation because third parties primarily experience impacts of such degradation, without cost to the agents responsible for the damage. Such rights – both community based and individual–are critical institutions mediating the relationships between humans and the use of the environment. Traditionally, village commons – water sources, grazing grounds, local forests, fisheries, etc., have been protected by local communities from overexploitation through various norms, which may include penalties for disallowed behaviour. These norms, may, however, be degraded through the very process of development, including urbanization, and population growth resulting from sharp reductions in mortality, and also through state actions which may create conditions for the strengthening of individual over communitarian rights and in doing so allow market forces to press for change that has adverse environmental implications. If such access to the community resources under weakened norms continue the resources would be degraded, and the livelihoods of the community would suffer.

Policy failures can emerge from various sources, including the use of fiscal instruments, such as explicit and implicit subsidies for the use of various resources, which provide incentives for excessive use of natural resources. Inappropriate policy can also lead to changes in commonly managed systems, with adverse environmental outcomes.

Another major set of challenges arise from emerging global environmental concerns such as climate change, stratospheric ozone depletion, and biodiversity loss. The key is to operationalize the principle of common but differentiated responsibility of countries in relation to these problems. Multilateral regimes and programmes responding to these global environmental issues must not adversely impact the development opportunities of developing countries. Further, the sharing of global natural resources must proceed only on the basis of equal sharing per-capita across all countries.

The causes, proximate and deeper, of degradation of key environmental resources are discussed below.

The Objectives of NEP 2004

The principal objectives of this policy are enumerated below. These objectives relate to current perceptions of key environmental challenges. They may, accordingly, evolve over time:

1. *Conservation of Critical Environmental Resources*: To protect and conserve critical ecological systems and resources, and invaluable natural and man-made heritage which are essential for lifesupport, livelihoods, economic growth, and a broad conception of human well-being.

2. *Intra-generational Equity*: Livelihood Security for the Poor: To ensure equitable access to environmental resources and quality for all sections of society, and in particular, to ensure that poor communities, which are most dependent on environmental resources for their livelihoods, are assured secure access to these resources.

3. *Inter-generational Equity*: To ensure judicious use of environmental resources to meet the needs and aspirations of present and future generations.

4. *Integration of Environmental Concerns in Economic and Social Development*: To integrate environmental concerns into policies, plans, programmes, and projects for economic and social development.

5. *Efficiency in Environmental Resource Use*: To ensure efficient use of environmental resources in the sense of reduction in their use per unit of economic output, to minimize adverse environmental impacts.

6. *Environmental Governance*: To apply the principles of good governance (transparency, rationality, accountability, reduction in time and costs, and participation) to the management and regulation of use of environmental resources.

7. *Enhancement of Resources for Environmental Conservation*: To ensure higher resource flows, comprising finance, technology, management skills, traditional knowledge, and social capital, for environmental conservation through mutually beneficial multistakeholder partnerships between local communities, public agencies, and investors.

Principles

The above objectives are to be realized through various strategic interventions by different public authorities at Central, State, and Local Government levels. They would also be the basis of partnerships between public agencies, local communities, and various economic actors. However, these strategic interventions, besides legislation and the evolution of legal doctrines for realization of the objectives, need to be premised on a core set of unambiguously stated principles. The following principles, accordingly, would guide the activities of different actors in relation to this policy. Each of these principles has an established genealogy in policy pronouncements, jurisprudence, international environmental law, or international State practice:

1. *Human beings are at the Centre of Sustainable Development Concerns*: Human beings are at the centre of concerns for sustainable development. They are entitled to a healthy and productive life in harmony with nature.

2. *The Right to Development*: The right to development must be fulfilled so as to equitably meet developmental and environmental needs of present and future generations.

3. *Environmental protection is an integral part of the development process*: In order to achieve sustainable development, environmental protection shall constitute an integral part of the development process and cannot be considered in isolation from it.

4. *The Precautionary Approach*: Where there are credible threats of serious or irreversible damage to key environmental resources, lack of full scientific certainty shall not be used as a reason for postponing cost-effective measures to prevent environmental degradation.

5. *Economic Efficiency*: In various public actions for environmental conservation, economic efficiency would be sought to be realized.

This principle requires that the services of environmental resources be given economic value, and such value to count equally with the economic values of other goods and services, in analysis of alternative courses of action.

Further implications of this principle are as follows:

"*Polluter Pays*": Impacts of acts of production and consumption of one party may be visited on third parties who do not have a direct economic nexus with the original act. Such impacts are termed "externalities". If the costs (or benefits) of the externalities are not re-visited on the party responsible for the original act, the resulting level of the entire sequence of production or consumption, and externality, is inefficient.

In such a situation, economic efficiency may be restored by making the perpetrator of the externality bear the cost (or benefit) of the same.

The policy will, accordingly, promote the internalisation of environmental costs, including through the use of incentives based policy instruments, taking into account the approach that the polluter should, in principle, bear the cost of pollution, with due regard to the public interest and without distorting international trade and investment.

Cost Minimization: Where the environmental benefits of a course of action cannot, for methodological or conceptual reasons, be imputed economic value (as in the case of "Incomparable Entities"), in any event the economic costs of realizing the benefits should be minimized.

Efficiency of resource use may also be accomplished by the use of policy instruments that create incentives to minimise wasteful use and consumption of natural resources. The principle of efficiency also applies to issues of environmental governance by streamlining processes and procedures in order to minimize costs and delays.

Entities with "Incomparable" Values: Significant risks to human health, life, and environmental life-support systems, besides certain other unique natural and man-made entities, which may impact the well-being broadly conceived of large numbers of persons, may be considered as "Incomparable" in that individuals or societies would not accept these risks for compensation in money or conventional goods and services. A conventional economic cost-benefit calculus would not, accordingly, apply in their case, and such entities would have priority in allocation of societal resources for their conservation without consideration of direct or immediate economic benefit.

Equity

The cardinal principle of equity or justice requires that human beings cannot be treated differently based on irrelevant differences between them. Equity norms must be distinguished according to context, *i.e.*, "procedural equity", relating to fair rules for allocation of entitlements and obligations, and "end result equity", relating to fair outcomes in terms of distribution of entitlements and obligations. Each context, in addition, must be distinguished in terms of "intra-generational equity", relating to justice within societies and in particular providing space for the participation of underprivileged men and women, and "inter-generational equity", relating to justice between generations.

Equity, in the context of this policy refers to both equity in entitlements to, and participation of the relevant publics in processes of decision-making over use of, environmental resources.

Legal Liability

Civil liability for environmental damage would deter environmentally harmful actions, and compensate the victims of environmental damage.

Conceptually, the principle of legal liability may be viewed as an embodiment in legal doctrine of the "polluter pays" approach, itself deriving from the principle of economic efficiency.

The following alternative approaches to legal liability may apply:

1. *Fault based liability*: In a fault based liability regime a party is held liable if it breaches a preexisting legal duty, for example, an environmental standard.

2. *Strict liability*: Strict liability imposes an obligation to compensate the victim for harm resulting from actions or failure to take action, which may not necessarily constitute a breach of any law or duty of care.

Public Trust Doctrine

The State is not an absolute owner, but merely a trustee of all natural resources, which are by nature meant for public use and enjoyment, subject to reasonable conditions, necessary to protect the legitimate interest of a large number of people, or for matters of strategic national interest.

Decentralisation

Decentralization involves ceding or transfer of power from a Central Authority to State and Local Authorities, in order to empower public authorities having jurisdiction at the spatial level at which particular environmental issues are salient, to address these issues.

Integration

Integration refers to the inclusion of environmental considerations in sectoral policymaking, the integration of the social and natural sciences in environment related policy research, and the strengthening of relevant linkages among various agencies at the Central, State, and Local Self- Government, charged with the implementation of environmental policies.

Environmental Standard Setting

Environmental standards must reflect the economic and social development situation in which they apply. Standards adopted in one society or context may have unacceptable economic and social costs if applied without discrimination in another society or context.

Setting en vironmental standards would involve several considerations, *i.e.*, risks to human health, risks to other environmental entities, technical feasibility, costs of compliance, and strategic considerations.

Preventive Action

It is preferable to prevent environmental damage from occurring in the first place, rather than attempting to restore degraded environmental resources after the fact.

Environmental Offsetting

There is a general obligation to protect threatened or endangered species and natural systems that are of special importance to sustaining life, providing livelihoods, or general well-being. If for exceptional reasons of overriding public interest such protection cannot be provided in particular cases, cost-effective offsetting measures must be undertaken by the proponents of the activity to restore as nearly as may be feasible the lost environmental services to the same publics.

Strategies and Actions

The foregoing statement of policy objectives and principles are to be realized by concrete actions in different areas relating to key environmental challenges. A large number of such actions are currently under way, and have been for several years, in some cases, for many decades. In some aspects new themes would need to be pursued to realize the principles and objectives. The following strategic themes, and outlines of actions to be taken in each, focus on both ongoing activities, functions, and roles, as well as new initiatives that are necessary. However, they are not necessarily a complete enumeration in each case.

Regulatory Reforms

The regulatory regimes for environmental conservation comprises a legislative framework, and a set of regulatory institutions. Inadequacies in each have resulted in accelerated environmental degradation on the one hand, and long delays and high transactions costs in development projects on the other. Apart from the legislation which is categorically premised on environmental conservation, a host of sectoral and cross-sector a l laws and policies, including fiscal regimes, also impact environmental quality (some of these are discussed in the succeeding sections).

Revisiting the Legislative Framework

The present legislative framework is broadly contained in the umbrella Environment Protection Act, 1986, the Water (Prevention and Control of Pollution) Act, 1974, the Water Cess Act, 1977 and the Air (Prevention and Control of Pollution) Act, 1981. The law in respect of management of forests and biodiversity is contained in the Indian Forest Act, 1927, the Forest (Conservation) Act 1980, the Wild Life (Protection) Act, 1972 and the Biodiversity Act, 2003. There are several other enactments, which complement the provisions of these basic enactments.

The following specific actions would be taken:

1. Institutionalize a holistic and integrated approach to the management of environment and natural resources, explicitly identifying and integrating environmental concerns in relevant sectoral and cross-sectoral policies through review and consultation, in line with the NEP, 2004.
2. Identify emerging areas for new legislation, due to better scientific understanding, economic and social development, and development of multilateral environmental regimes, in line with NEP, 2004.

3. Review the body of existing legislation in order to develop synergies among relevant statutes and regulations, eliminate obsolescence, and amalgamate provisions with similar objectives, in line with NEP, 2004.

4. Ensure accountability of the concerned levels of Government (Centre, State, Local) in undertaking the necessary legislative changes in a defined time-frame, with due regard to the Objectives and Principles of NEP, 2004, in particular, ensuring the livelihood and wellbeing of the poor.

Process Related Reforms

Approach

The recommendations of the Committee on Reforming Investment Approval and Implementation Procedures (The Govindarajan Committee identified delays in environment and forest clearances as the largest source of delays in development projects–Appendix I), will be followed for reviewing the existing procedures for granting clearances and other approvals under various statutes and rules. These include the Environment Protection Act, Forest Conservation Act, the Water (Prevention and Control of Pollution) Act, the Air (Prevention and Control of Pollution) Act and Wildlife (Protection) Act, and Genetic Engineering Approval Committee (GEAC) Rules under the Environment Protection Act. The objective is to reduce delays and levels of decision-making, realiz e decentralization of environmental functions, and ensure greater transparency and accountability.

Framework for Legal Action

The present approach to dealing with environmentally unacceptable behaviour in India has been largely based on criminal processes and sanctions. Although criminal sanctions, if successful, may create a deterrent impact, in reality they are rarely fruitful for a number of reasons. On the other hand, giving lower level officials the power to institute criminal prosecutions may provide fertile opportunities for rent-seeking.

Civil law, on the other hand, offers flexibility, and its sanctions can be more effectively tailored to particular situations. The evidentiary burdens of civil proceedings are less daunting than those of criminal law. It also allows for preventive policing through orders and injunctions to restrain prospective pollution.

Accordingly, a judicious mix of civil and criminal processes and sanctions will be employed in the legal regime for enforcement, through a review of the existing legislation. Civil liability law, civil sanctions, and processes would govern most situations of non-compliance. Criminal processes and sanctions would be available for serious, and potentially provable, infringements of environmental law, and their initiation would be vested in responsible authorities. Recourse may also be had to the relevant provisions in the Indian Penal Code, and the Criminal Procedure Code.

Substantive Reforms

Environment and Forests clearances In order to make the clearance processes more effective, the following actions will be taken:

1. Encourage regulatory authorities, Central and State, to institutionalise regional and cumulative environmental impact assessments (R/CEIAs) to ensure that environmental concerns are identified and addressed at the planning stage itself.

2. Give due consideration, to the quality and productivity of lands which are proposed to be converted for development activities, as part of the clearance process. Projects involving large-scale diversion of prime agricultural land would require environmental clearance whether or not the proposed activity otherwise requires environmental clearance.

3. Encourage clustering of industries and other development activities to facilitate setting up of environmental management infrastructure, as well as monitoring and enforcing environmental compliance. Emphasize post project monitoring and implementation of environmental management plans through participatory processes, involving the government, industry, and the potentially impacted community.

4. Prohibit the diversion of dense natural forests to non-forest use, except in site -specific cases of vital national interest. No further regularisation of encroachment on forests should be permitted.

Coastal Areas

Development activities in the coastal areas are regulated by means of the Coastal Regulation Zone notifications and Integrated Coastal Zone Management (ICZM) Plans made under them. However, there is need to ensure that the regulations are firmly founded on scientific principles, in order to ensure effective protection to valuable coastal environmental resources, without unnecessarily impeding livelihoods, or legitimate coastal economic activity, or settlements, or infrastructure development.

The following actions would be taken:

1. Revisit the Coastal Regulation Zone (CRZ) notifications to make the approach to coastal environmental regulation more holistic, and thereby ensure protection to coastal ecological systems, coastal waters, and the vulnerability of some coastal areas to potential sea level rise. The Integrated Coastal Zone Management (ICZM) Plans need to be comprehensive, and prepared on scientific basis, with the participation of the local communities both in formulation and implementation. The ICZM Plans should be reviewed at pre-determined intervals to take account of changes in geomorphology, economies, and settlement patterns.

2. Decentralize, to the extent feasible, the clearance of specific projects to State environmental authorities, exempting activities, which do not cause significant environmental impacts, and are consistent with approved ICZM Plans.

Living Modified Organisms (LMOs)

Biotechnology has an immense potential to enhance livelihoods and contribute to the economic development of the country. On the other hand, LMOs may pose significant risks to ecological resources, and perhaps, human and animal health. In order to ensure that development of biotechnology does not lead to unforeseen adverse impacts, the following actions will be taken:

Review the regulatory processes for LMOs so that all relevant scientific knowledge is taken into account, and ecological, health, and economic concerns are adequately addressed.

Periodically review the National Bio-safety guidelines and Bio-safety Operations Manual to ensure that these are based on current scientific knowledge.

Ensure the conservation of bio-diversity and human health when dealing with LMOs in transboundary movement in a manner consistent with the Multilateral Bio-safety Protocol.

Environmentally Sensitive Zones

Environmentally Sensitive Zones may be defined as areas with identified environmental resources with "Incomparable Values " which require special attention for their conservation. In order to conserve and enhance these resources, without impeding legitimate socio-economic development of these areas, the following actions will be taken:

Identify and give legal status to Environmentally Sensitive Zones in the country with environmental entities with "Incomparable values" requiring special conservation efforts.

Formulate area development plans for these zones on a scientific basis, with adequate participation by the local communities.

Create local institutions with adequate participation for the environmental management of such areas to ensure adherence to the approved area development plans, which should be prepared in consultation with the local communities.

Monitoring and Enforcement

Weak enforcement of environmental compliance is attributed to inadequate technical capacities, monitoring infrastructure, and trained staff in enforcement institutions. In addition, there is insufficient involvement of the potentially impacted local communities in the monitoring of compliance, and absence of institutionalised public-private partnerships in enhancement of monitoring infrastructure.

The following actions would be taken:

Give greater legal standing to local community based organizations to undertake monitoring of environmental compliance, and report violations to the concerned enforcement authorities.

Develop feasible models of public-private partnerships to leverage financial, technical, and management resources of the private sector in setting up and operating infrastructure for monitoring of environmental compliance, with ironclad safeguards against possible conflict of interest or collusion with the monitored entities.

Use of economic principles in environmental decision-making: It is necessary that the costs associated with the degradation and depletion of natural resources be incorporated into the decisions of economic actors at various levels to reverse the tendency to treat these resources as "free goods" and to pass the costs of degradation to other sections of society, or to future generations of the country.

At the macro-level, a system of natural resource accounting is required to assess whether in the course of economic growth we are drawing down, or enhancing, the natural resource base of production, including all relevant depletable assets. In addition, the environmental costs and benefits associated with various activities, including sectoral policies, should be evaluated to ensure that these factors are duly taken into account in decision-making.

The current near exclusive reliance on fiats based instruments for environmental regulation do not permit individual actors to minimize their own costs of compliance. This leads, on the one hand, to non-compliance in many cases, and on the other, unnecessary diversion of societal resources from other pressing needs. Economic instruments, of which a large, feasible suite has emerged through practical experience in several developed and developing countries, work by aligning the interests of economic actors with environmental compliance, primarily through application of "polluter pays". This may ensure that for any given level of environmental quality desired, the society-wide costs of meeting the standard are minimized. However, in some cases, use of economic instruments may require intensive monitoring, which too may entail significant societal costs. On the other hand, use of existing policy instruments, such as the fiscal regime, may significantly reduce or eliminate the need for enhanced institutional capacities to administer the incentive based instruments. In future, accordingly, a judicious mix of incentives based and fiats based regulatory instruments would be considered for each specific regulatory situation.

The following actions would be taken:

1. Strengthen the initiatives being taken by the Central Statistical Organization in the area of natural resource accounting with a view to its adoption in the system of national income accounts.

2. Develop and promote the use of standardized environmental accounting practices and standards in preparation of statutory financial statements for large industrial enterprises, in order to encourage greater environmental responsibility in investment decision-making, management practices, and public scrutiny.

3. Encourage financial institutions to adopt appraisal practices, so that environmental risks are adequately considered in the financing of projects.

4. Facilitate the integration of environmental values into cost-benefit analysis to encourage more efficient allocation of resources while making public investment and policy decisions.

5. Prepare and implement an action plan on the use of economic instruments for environmental regulation in specified contexts.

Enhancing and Conserving Environmental Resources

Perverse production and consumption practices are the immediate causes of environmental degradation, but an exclusive focus on these aspects alone is insufficient to prevent environmental harm. The causes of degradation of environmental resources lie ultimately in a broad range of policy, and institutional, including regulatory shortcomings, leading to the direct causes. However, the range of policies, and legal and institutional regimes, which impact the proximate factors, is extremely wide, comprising fiscal and pricing regimes, and sectoral and cross - sectoral policies, laws, and institutions. Accordingly, apart from programmatic approaches, review and reform of these regimes to account for their environmental consequences is essential. In addition, there is lack of awareness of the causes and effects of environmental degradation, and how they may be prevented, among both specialized practitioners of the relevant professions, including policymakers, as well as the general public, which needs to be redressed. In this subsection, in respect of major categories of environmental resources, the proximate and deeper causes of their degradation, and specific initiatives for addressing them are outlined.

Land Degradation

The degradation of land, through soil erosion, alkali-salinization, waterlogging, pollution, and reduction in organic matter content has several proximate and underlying causes. The proximate causes include loss of forest and tree cover (leading to erosion by surface water run-off and winds), excessive use of irrigation (in many cases without proper drainage, leading to leaching of sodium and potassium salts), improper use of agricultural chemicals (leading to accumulation of toxic chemicals in the soil), diversion of animal wastes for domestic fuel (leading to reduction in soil nitrogen and organic matter), and disposal of industrial and domestic wastes on productive land. These in turn, are driven by implicit and explicit subsidies for water, power, fertilizer and pesticides, and absence of conducive policies and regulatory systems to enhance people's incentives for afforestation and forest conservation. It is essential that the relevant fiscal, tariffs, and sectoral policies take explicit account of their unintentional impacts on land degradation, if the fundamental basis of livelihoods for the vast majority of our people is not to be irreparably damaged. In addition, to such policy review, the following specific initiatives would be taken:

1. Encourage adoption of science-based, and traditional sustainable land use practices through research and development, pilot scale demonstrations, and large scale dissemination, including farmer's training, and where necessary, access to institutional finance.

2. Promote reclamation of wasteland and degraded forestland through formulation and adoption of multistakeholder partnerships involving the land owning agency, local communities, and investors.

3. Prepare and implement thematic action plans for arresting and reversing desertification.

Forests and Wildlife

Forests

Forests provide a multiplicity of environmental services. Foremost among these is the recharging of mountain aquifers, which sustain our rivers. They also conserve the soil, and prevent floods and drought. They provide habitat for wildlife and the ecological conditions for maintenance and natural evolution of genetic diversity of flora and fauna. They are the traditional homes of forest dwelling tribals, the major part by far of whose livelihoods depend on forests. They yield timber, fuel-wood, and other forest produce, and possess immense potential for economic benefits, in particular for local communities, from sustainable eco-tourism.

On the other hand, in recent decades, there has been significant loss of forest cover, although there are now tangible signs of reversal of this trend. The principal direct cause of forest loss has been the conversion of forests for agriculture, settlements, infrastructure, and industry. In addition, commercial extraction of fuel-wood, illegal felling, and grazing of cattle, has degraded forests. These causes, however, have their origins in the fact that the environmental values provided by forests are not realized as direct financial benefits by various parties, at least to the extent of exceeding the monetary incomes from alternative uses, including those arising from illegal use. Moreover, while since antiquity forest dwelling tribes had generally recognized traditional community rights over the forests, on account of which they had strong incentives to use the forests sustainably and to protect them from encroachers, following the commencement of formal forest laws and institutions in 1865, these rights were effectively extinguished in many parts of the country. Such disempowerment has led to the forests becoming open access in nature, leading to their gradual degradation in a classic manifestation of the "*Tragedy of the Commons*", besides leading to perennial conflict between the tribals and the Forest Department, and constituting a major denial of justice.

It is possible that some site-specific non-forest activities may yield overall societal benefits significantly exceeding that from the environmental services provided by the particular tract of forest. However, large scale forest loss would lead to catastrophic, permanent change in the country's ecology, leading to major stress on water resources and soil erosion, with consequent loss of agricultural productivity, industrial potential, living conditions, and the onset of natural disasters including drought and floods. In any event, the environmental values of converted forests must be restored, as nearly as may be feasible, to the same publics.

The National Forest Policy, 1988, and the Indian Forest Act, as well as the regulations under it, provide a comprehensive basis for forest conservation. However, it is necessary, looking to some of the underlying causes of forest loss, to take some further steps. These include:

1. Give legal recognition of the traditional rights of forest dwelling tribes. This would remedy a serious historical injustice, secure their livelihoods, reduce possibilities of conflict with the Forest Departments, and provide long-term incentives to the tribals to conserve the forests.

2. Formulate an innovative strategy for increase of forest and tree cover from the present level of 23 per cent of the country's land area, to 33 per cent in 2012, through afforestation of degraded forest land, wastelands, and tree cover on private or revenue land. Key elements of the strategy would include: (i) the implementation of multistakeholder partnerships involving the Forest Department, local communities, and investors, with clearly defined obligations and entitlements for each partner, following good governance principles, to derive environmental, livelihood, and financial benefits; (ii) rationalization of restrictions on cultivation of forest species outside notified forests, to enable farmers to undertake social and farm forestry where their returns are more favourable than cropping, and (iii) universalization of the Joint Forestry Management (JFM) system throughout the country.

3. Focus public investments on enhancing the density of natural forests, mangroves conservation, and universalization of Joint Forestry Management.

4. Formulate an appropriate methodology for reckoning and restoring the environmental values of forests, which are unavoidably diverted to other uses.

5. Formulate and implement a "Code of Best Management Practices" for dense natural forests to realize the Objectives and Principles of NEP, 2004.

Wildlife

The status of wildlife in a region is an accurate index of the state of ecological resources, and thus of the natural resource base of human well being. This is because of the interdependent nature of ecological entities (*"the web of life"*), in which wildlife is a vital link. Moreover, several charismatic species of wildlife embody "Incomparable Values", and at the same time, are a major resource base for sustainable eco-tourism.

Conservation of wildlife, accordingly, involves the protection of entire ecosystems. However, in several cases, delineation of and restricting access to such Protected Areas (PAs), as well as encroachment of human settlements on these areas has led to man-animal conflicts. While physical barriers may temporarily reduce such conflict, it is preferable to address their underlying causes. These may largely arise from the non-involvement of relevant stakeholders in identification and delineation of PAs.

In respect of Wildlife Conservation, the following elements would be pursued:

1. Expand the Protected Area (PA) network of the country, including Conservation and Community Reserves, to give fair representation to all biogeographic zones of the country. In doing so, develop norms for delineation in terms of the Objectives and Principles of NEP, 2004, in particular, participation of local communities, concerned public agencies, and other stakeholders, to harmonize ecological and physical features with needs of socio -economic development. It must be ensured that the overall are a of the network, in each bio-geographic zone would increase in the process.

2. Paralleling multistakeholder partnerships for afforestation, formulate and implement similar partnerships for enhancement of wildlife habitat in Conservation Reserves and Community Reserves, to derive both environmental and eco-tourism benefits.

3. Promote site-specific eco-development programmes in fringe areas of PAs, to restore livelihoods and access to forest produce by local communities owing to access restrictions in PAs.

4. Strengthen capacities and implement measures for captive breeding and release into the wild identified endangered species.

Biodiversity, Traditional Knowledge, and Natural Heritage

Biodiversity, comprises both genetic and ecosystems diversity. Loss of biodiversity is primarily due to degradation or alteration of ecosystems, in particular the habitats of site-specific species. Damage to such habitats arises from land degradation, forest loss, conversion of wetlands, pollution of and excessive water drawals from rivers, and loss of coastal ecosystems, the reasons for which have been discussed separately. Conservation of genetic diversity, in particular, is crucial for development of improved crop varieties resistant to particular stresses, new pharma products, etc., apart from ensuring the resilience of ecosystems. However, it is presently difficult to foresee the future potential of any particular genetic resource, and accordingly economic values are uncertain. Traditional Knowledge (TK), referring to ethno-biology knowledge possessed by local communities, relates to uses of various indigenous plant and faunal varieties, including in traditional medicine, food, etc., and is potentially an important means of unlocking the value of genetic diversity through reduction in search costs.

Natural heritage sites, including endemic "biodiversity hotspots", sacred groves and landscapes, are repositories of significant genetic and ecosystem diversity, and the latter are also important bases for eco-tourism. They are nature's laboratories for evolution of wild species in response to change in environmental conditions.

India is fortunate in having, through the efforts of dedicated scientists over many decades, developed vast inventories of floral and faunal resources, as well as ethno-biology knowledge. India is, thus well-placed to tap this enormous resource base for benefits for the country as a whole, and local communities in particular, provided that the genetic resources are conserved, and appropriate Intellectual Property Rights (IPRs) conferred on local communities in respect of their ethno-biology knowledge.

A large-scale exercise has been completed for providing inputs towards a National Biodiversity Action Plan. These inputs would be reviewed in terms of the Objectives and Principles of NEP, 2004, scientific validity, financial and administrative feasibility, and legal aspects. In any event, the following measures would be taken:

1. Strengthen the protection of areas of high endemism of genetic resources ("biodiversity hot spots"), while providing alternative livelihoods and access to resources to local communities who may be affected thereby.

2. Pay explicit attention to the potential impacts of development projects on biodiversity resources and natural heritage. In appraisal of such projects by cost-benefit analysis, assign values to biodiversity resources at or near the upper end of the range of uncertainty. In particular, ancient sacred groves and "biodiversity hotspots" should be treated as possessing "Incomparable Values".

3. Enhance *ex-situ* conservation of genetic resources in designated gene banks across the country. Genetic material of threatened species of flora and fauna must be conserved on priority.

4. Formulate and adopt an internationally recognized system of legally enforceable *sui-generis* intellectual property rights for the country's genetic resources, to enable the country, including where relevant the local communities, to derive economic benefits from grant of access to these resources.

5. Similarly, formulate and adopt an internationally recognized system of legally enforceable *sui-generis* intellectual property rights for ethno-biology knowledge, to enable local communities to realize significant financial benefits from permitting the use of such knowledge. Set up an online database of the inventory of such ethno-biology knowledge, once the legal regime, domestic and multilateral, for their protection is in place.

Freshwater Resources

India's freshwater resources comprise the single most important class of natural endowments enabling its economy and its human settlement patterns. The freshwater resources comprise the river systems, groundwater, and wetlands. Each of these has a unique role, and characteristic linkages to other environmental entities.

River Systems

India's river systems typically originate in its mountain eco-systems, and deliver the major part of their water resources to the populations in the plains. They are subject to siltation from sediment loads due to soil loss, itself linked to loss of forest and tree cover. They are also subject to significant net water withdrawals along their course, due to agricultural, industrial, and municipal use; as well as pollution from human and animal waste, agricultural run-offs, and industrial effluents. Although the rivers possess significant natural capacity to assimilate and render harmless many pollutants, the existing pollution inflows in most cases substantially exceed such natural capacities. This fact, together with progressive reductions in stream flows, ensures that the river water quality in the vast majority of cases declines as one goes downstream. The results include loss of aquatic flora and fauna, leading to loss of livelihoods for river fisherfolk, significant impacts on human health from polluted water, loss of habitat for many bird species, and loss of inland navigation potential. Apart from these, India's rivers are inextricably linked with the history and religious beliefs of its peoples, and the degradation of important river systems accordingly offends their spiritual, aesthetic, and cultural sensibilities.

The broad direct causes of rivers degradation are, in turn, linked to several policies and regulatory regimes. These include tariff policies for irrigation systems and industrial use, which, through inadequate cost-recovery, provide incentives for overuse near the headwork's of irrigation systems, and drying up of irrigation systems at the tail-ends. The result is excessive cultivation of water intensive crops near the headwork's, which is otherwise inefficient, waterlogging, and alkali-salinization of soil. The irrigation tariffs also do not yield resources for proper maintenance of irrigation systems, leading to loss in their potential; in particular, resources are generally not available for lining irrigation canals to prevent seepage loss. These factors result in reduced flows in the rivers. Pollution loads are similarly linked to pricing policies leading to inefficient use of agricultural chemicals, and municipal and industrial water use. In particular, revenue yields for the latter two are insufficient to install and maintain sewage and effluent treatment plants, respectively. Pollution regulation for industries is typically not based on formal spatial planning to facilitate clustering of industries to realize scale economies in effluent treatment, resulting in relatively high costs of effluent treatment, and consequent increased incentives for non-compliance. There is, accordingly need to review the relevant pricing policy regimes and regulatory mechanisms in terms of their likely adverse environmental impacts.

The following comprise elements of an action plan for river systems:

1. Promote integrated approaches to management of river basins by the concerned river authorities, considering upstream and downstream inflows and withdrawals by season, pollution loads and natural regeneration capacities, to ensure maintenance of adequate flows and adherence to water quality standards throughout their course in all seasons.
2. Consider and mitigate the impacts on river flora and fauna, and the resulting change in the resource base for livelihoods, of multipurpose river valley projects, power plants, and industries.
3. Consider mandating the installation of water saving closets and taps in the building byelaws of urban centres.

Groundwater

Groundwater is present in underground aquifers in many parts of the country. Aquifers near the surface are subject to annual recharge from precipitation, but the rate of recharge is impacted by human interference. Deep aquifers, on the other hand, occur below a substratum of hard rock. The deep aquifers generally contain very pure water, but since they are recharged only over many millennia, must be conserved for use only in periods of calamitous drought such as may happen only once in several hundred years. The boundaries of groundwater aquifers do not generally correspond to the spatial jurisdiction of any local public authorities or private holdings, nor are they easily discernable, nor can withdrawals be easily monitored, leading to the unavoidable situation of groundwater being an open access resource.

The water table has been falling rapidly in many areas of the country in recent decades. This is largely due to withdrawal for agricultural, industrial, and urban use, in excess of annual recharge. In urban areas, apart from withdrawals for domestic and industrial use, housing and infrastructure such as roads, prevent sufficient recharge. In addition, some pollution of groundwater occurs due to leaching of stored hazardous waste and use of agricultural chemicals, in particular, pesticides. Contamination of groundwater is also due to geogenic causes, such as leaching of arsenic from natural deposits. Since groundwater is frequently a source of drinking water, its pollution leads to serious health impacts.

The direct causes of groundwater depletion have their origin in the pricing policies for electricity and diesel. In the case of electricity, where individual metering is not practiced, a flat charge for electricity connections makes the marginal cost of electricity effectively zero. Subsidies for diesel also reduce the marginal cost of extraction to well below the efficient level. Given the fact that groundwater is an open access resource, the user then "rationally" (*i.e.*, in terms of his individual perspective), extracts groundwater until the marginal value to him equals his now very low marginal cost of extraction. The result is inefficient withdrawals of groundwater by all users, leading to the situation of falling water tables. Support prices for several water intensive crops with implicit price subsidies aggravate this outcome by strengthening incentives to take up these crops rather than less water intensive ones.

Falling water tables have several perverse social impacts, apart from the likelihood of mining of deep aquifers, *"the drinking water source of last resort"*. The capital costs of pump sets and bore wells for groundwater extraction when water tables are very deep may be relatively high, with no assurance that water would actually be found. In such a situation, a user who may be a marginal farmer able to borrow the money only at usurious rates of interest, may, in case water is not found, find it impossible to repay his debts. This may lead to destitution, or worse. Even if the impacts were not so dire, there would be excessive use of electricity and diesel.

The efficient use of groundwater would, accordingly, require that the practice of non-metering of electric supply to farmers be discontinued in their own enlightened self-interest. It would also be essential to progressively ensure that the environmental impacts are taken into account in setting electricity tariffs, and diesel pricing.

Increased run-off of precipitation in urban areas due to impermeable structures and infrastructure prevents groundwater recharge. This is an additional cause of falling water tables in urban areas. In rural areas several cost-effective contour bunding techniques have been proven to enhance groundwater recharge. A number of effective traditional water management techniques to recharge groundwater have been discontinued by the local communities due to the onset of pump sets extraction, and need to be revived. Finally, increase in tree cover, is also effective in enhancing groundwater recharge.

Chapter 15
Ability-to-Pay Principle

A principle of taxation in which taxes are based on the income or resource-ownership ability of people to pay the tax. The income tax is one of the most common taxes that seeks to abide by the ability-to-pay principle. In theory, the income tax system is set up such that people with greater incomes pay more taxes. Proportional and progressive taxes follow this ability-to-pay principle, while regressive taxes, such as sales taxes and Social Security taxes, don't.

The logic behind the ability-to-pay principle is that taxes are collected by the government to finance public goods that provide benefits to all members of society. And because taxes are a diversion of resources from the household to the government sector, it makes sense to tax, or divert income away from, the people who actually have the income.

Absolute Advantage: The ability of a producer to produce a higher absolute quantity of a good with the productive resource available.

Abundance: A term that applies when individuals can obtain all the goods they want without cost. If a good is abundant, it is free.

Accelerator: The causal relationship between changes in consumption and changes in investment.

Acid Rain: The precipitation of dilute solutions of strong mineral acids, formed by the mixing in the atmosphere of various industrial pollutants – primarily sulphur dioxide and nitrogen oxides – with naturally occurring oxygen and water vapor.

Aquifer: Underground source of water

Acquired Endowments: Resources a country builds for itself, like a network of roads or an educated population

Adaptive Expectations: Expectations based on the extrapolation of events in the recent past into the future

Adverse Selection: Principle that says that those who most want to buy insurance tend to be those most at risk, but charging a high price for insurance (to cover the high risk) will discourage those at less risk from buying insurance at all

When a negotiation between two people with asymmetric information restricts the quality of the good traded. This typically happens because the person with more information can negotiate a favorable exchange. This is frequently referred to as the "market for lemons."

For example, let's say you're searching for a car, knowing that some are "high-quality" and others are "low-quality." However, you don't know which category a particular car is in. Suppose there's an equal chance of getting either a high-quality or low-quality car. If you're willing to pay $2,000 for a high quality car, but only $1,000 for the low quality car, how much would you offer for a given car of unknown quality?

The expected value of the car is $1,500. In other words, if you bought hundreds of cars, half worth $2,000 and half worth $1,000, the average value of the cars is $1,500 each. Not knowing the quality of a given car, the price you would offer is $1,500- the average or expected price. The chance of overpaying for a low-quality car is offset by the chance of underpaying for a high-quality car.

Unlike you, each owner is better aware of the quality of his or her car – they have more information than you. Your $1,500 offer would be accepted by the seller of a low-quality car, but refused by the seller of the high-quality car. Due to the lack of buyers' information, high-quality cars would not be sold. The only cars exchanged would be low-quality cars ("lemons"). Asymmetric information tends to limit quality of products exchanged, adversely selecting the lower quality cars.

Aggregate Demand Curve: A curve relating the total demand for the economy's goods and services at each price level, given the level of wages

Aggregate Expenditures Schedule: A curve that traces out the relationship between expenditures–the sum of consumption, investment, government expenditures, and net exports–and the national income, at a fixed price level

Aggregate Supply Curve: A curve relating the total supply of the economy's goods and services at each price level, given the level of wages.

Allocative Efficiency: Obtaining the most consumer satisfaction from available resources

Ambient Charge: A form of tax on non uniformly mixed pollutants. It is calculated to be the same in terms of the emission's impact on ambient environmental quality at some receptor site. As a result, an ambient charge to a firm closer to the receptor site will normally be higher *per litre* than that charged to firms further away.

Antitrust Laws: Designed to promote open markets by limiting practices that reduce competition.

Assets: Any item that is long-lived, purchased for the service it renders over its life and for what one will receive when one sells it.

What a person or business owns.

Assistance In Kind: Public assistance that provides particular goods and services, like food or medical care, rather than cash

Asymmetric Information: A situation in which the parties to a transaction have different information, as when the seller or a used car has more information about its quality then the buyer.

The economics of information search tells us that everyone falls short of having perfect information. It suggests that everyone will have different information about different things. For example, if you aren't a plumber (nor have any desire to become one), then you aren't likely to seek information about the wages paid plumbers in Boise, Idaho. In contrast, this information could be quite beneficial to plumbers in Pocatello, Idaho.

Asymmetric Information for the market occurs when buyers and sellers have different information about a good. Sellers often have better information about a good than buyers because they are more familiar with it. They know more about it's quality, durability, and other features. Buyers, in contrast, have limited contact with the commodity and thus have less information.

For example, if you sell a car that you've owned for several years, you know how well it's been maintained, whether or not it needs frequent repairs, and what causes that strange "clanking" sound. A buyer who test drives the car for only a few miles is likely unaware of these facts. And because search cost is positive, the buyer is unlikely to acquire as much information about the car as you already possess.

Another common example of asymmetric Information is in the labour market. Workers are knowledgeable about their skills, industriousness, and productivity. Employers, in contrast, have limited information about the quality of prospective workers.

Average Costs: The total costs divided by the total output.

Average Productivity: Total quantity divided by the total quantity of input.

Average Variable Costs: The total variable costs divided by the total output.

Balance of Payments: A record of all the financial transactions between a country and the rest of the world during a given year.

Barriers To Entry: Factors that prevent firms from entering a market, such as government rules or patents.

Basic Competitive Model: The model of the economy that pulls together the assumptions of self-interested consumers, profit maximizing firms, and perfectly competitive markets.

Benefit-Cost Analysis: A tally/comparison of expenditures and advantages in dollar terms resulting from various actions.

Benefits in Kind: Non-cash forms of pay or assistance.

Bequest Savings Motive: People save so that they can leave an inheritance to their children.

Bequest Values: Willingness to pay to preserve the environment for the benefit of our children and grandchildren.

Bertrand Competition: An oligopoly in which each firm believes that its rivals are committed to keeping their prices fixed and that customers can be lured away by offering lower prices.

Brownfields: Abandoned, idled, or under-used industrial and commercial facilities where expansion or redevelopment is complicated by real or perceived environmental contamination.

Business Cycles: Periodic swings in the pace of national economic activity, characterized by alternating expansion and contraction phases.

Capital: The existing stock of productive resources, such as machines and buildings, that have been produced. Capital Intensive: Production methods with a high quantity of capital per worker.

Capital Gain: The increase in the value of an asset between the time it is purchased and the time it is sold.

Capital Market: The market in which savings are made available to investors.

Capitalist Economies: Economies which use market-determined prices to guide peoples choices about the production and distribution of goods; these economies generally have productive resource which are privately owned.

Carbon Tax: A charge on fossil fuels (coal, oil, natural gas) based on their carbon content. When burned, the carbon in these fuels becomes carbon dioxide in the atmosphere, the chief greenhouse gas.

Carcinogens: Substances that cause cancer.

Cartel: A group of producers with an agreement to collude in setting prices and output.

Categorical Assistance: Public assistance aimed at a particular category of people, like the elderly or the disabled.

Causation: Relationship that results when an change in one variable is not only correlated with but actually causes the change in another one.

Central Planning: The system in which central government bureaucrats (as opposed to private entrepreneurs or even local government bureaucrats) determine what will be produced an how it will be produced.

Centralization: Organizational structure in which decision making is concentrated at the top.

Centrally Planned Economy: An economy in which most decisions about resource allocation are made by the central government.

CERCLA: Comprehensive Environmental Response, Compensation and Liability Act. U.S. federal law enacted by Congress in 1980 for the purpose of cleaning up existing toxic sites. A.K.A. Superfund.

Change in Demand: A shift in the entire demand curve so that at any given price, people will want to buy a different amount. A change in demand is caused by some change other than a change in the goods price.

Change in Quantity Demanded: Movement up or down a given demand curve caused by a change in the goods price with no shift in the curve itself.

Change in Quantity Supplied: A price change causing movement along the supply curve but no shift in the position of the curve itself.

Change in Supply: A change in one of the cost determinants of supply causing a shift in the position of the supply curve.

Choice: The act of selecting among alternatives, a concept crucial to economics.

Chlorofluorocarbons (CFCs): Stable, artificially-created chemical compounds containing carbon, chlorine, fluorine and sometimes hydrogen. Chlorofluorocarbons, used primarily to facilitate cooling in refrigerators and air conditioners, have been found to damage the stratospheric ozone layer which protects the earth and its inhabitants from excessive ultraviolet radiation.

Civilian Labor Force: All persons over the age of sixteen who are not in the armed forces nor institutionalized and who are either employed or unemployed.

Classical Economists: Economists prevalent before the Great Depression who believed that the basic competitive model provided a good description of the economy and that if short periods of unemployment did occur, market forces would quickly restore the economy to full employment.

Classical Unemployment: Unemployment that results from too-high real wages; it occurs in the supply constrained equilibrium, so that rightwards shifts in aggregate supply reduce the level of unemployment.

Clean Fuel: Fuels which have lower emissions than conventional gasoline and diesel. Refers to alternative fuels as well as to reformulated gasoline and diesel.

Cleanup: Treatment, remediation, or destruction of contaminated material.

Clearcutting: A logging technique in which all trees are removed from an area, typically 20 acres or larger, with little regard for long-term forest health.

Climate Change: A regional change in temperature and weather patterns. Current science indicates a discernible link between climate change over the last century and human activity, specifically the burning of fossil fuels.

Closed Economy: An economy that neither exports nor imports.

Coase's Theorem: The assertion that if property rights are properly defined, then people will be forced to pay for any negative externalities they impose on others, and market transactions will produced efficient outcomes.

Common Property Resources: Resources for which there are no clearly defined property rights; property owned in common by a society.

Community Right-to-Know: Public accessibility to information about toxic pollution.

Compact Fluorescent: Flourescent light bulbs small enough to fit into standard light sockets, which are much more energy-efficient than standard incandescent bulbs.

Comparative Advantage: The ability of a producer to produce a good at a lower marginal cost than other producers; marginal cost in the sacrifice of some other good compared to the amount of a good obtained. A country has an comparative advantage over another in one good as opposed to another good if its relative efficiency in the production of the first good is higher than the other country's.

Compensating Variation: The amount of money one would pay to gain a benefit such as a price decrease or the amount of income one would accept to agree upon the imposition of a harm such as a price increase. Money required to leave an individual as well off as before the economic change. Amount an individual would be willing to pay for the change, or willing to accept as compensation for a change.

Compensating Wage Differentials: The additional amount paid for a job that has certain unattractive features, such as risk of injury, as compared with a job that requires similar skills but lacks these negative features.

Competition: Rivalry among individuals in order to acquire more of something that is scarce.

Competitive Equilibrium Price: The price at which the quantity supplied and the quantity demanded are equal to each other.

Complement: A good for which demand decreases when the price of a closely related good increases.

Complements: A price change for one product leads to a shift in the opposite direction in the demand for another product.

Compliance Costs: Expenditures associated with fulfilling requirements of environmental regulations.

Compost: Process whereby organic wastes, including food wastes, paper, and yard wastes, decompose naturally, resulting in a product rich in minerals and ideal for gardening and farming as a soil conditioners, mulch, resurfacing material, or landfill cover.

Comprehensive Environmental Response, Compensation and Liability Act: U.S. federal law enacted by Congress in 1980 for the purpose of cleaning up existing toxic sites. A.K.A. Superfund, CERCLA.

Consumer Price Index (CPI): A measure of the average amount (price) paid for a market basket of goods and services by a typical U.S. consumer in comparison to the average paid for the same basket in an earlier base year. A price index in which the basket of goods is defined by what a typical consumer purchases.

Consumer Protection Legislation: Laws aimed at protecting consumers, for instance by assuring that consumers have more complete information about items they are considering buying.

Consumer Sovereignty: The principle that holds that each individual is the best judge of what makes him better off.

Consumer Surplus: The difference between what a person would be willing to pay and what he actually has to pay to buy a certain amount of a good.

Consumption Expenditures: The total dollar value of all goods and services purchased by the household sector for current use.

Consumption Function: A mathematical expression relating personal consumption expenditures to disposable income. The relationship between disposable income and consumption.

Contingency Clauses: Statements within a contract that make the level of payment or the work to be performed conditional upon various factors.

Contingent Valuation Method: Directly asks people what they are willing to pay for a benefit an/or willing to receive in compensation for tolerating a cost through a survey or questionnaire. Personal valuations for increases or decreases in the quantity of some good are obtained contingent upon a hypothetical market. The aim is to elicit valuations or bids which are close to what would be revealed if an actual market existed. Several biases, including strategic, design, (starting point, vehicle, and informational), hypothetical, and operational are discussed above and below.

Corporate Income Tax: A tax based on the income, or profit, received by a corporation.

Constant Returns To Scale: When all inputs are increased by a certain proportion, output increases by the same proportion.

Correlation: Relationship that results when a change in one variable is consistently associated with a change in another one.

Cost: The most valuable opportunity forsaken when a choice is made.

Cost-Benefit Analysis: A tally/comparison of expenditures and advantages in dollar terms resulting from various actions.

Cost-Effectiveness Analysis: Least expensive way of achieving a given environmental quality target, or the way of achieving the greatest improvement in some environmental target for a given expenditure of resources.

Cost-of-Living Adjustments: Automatic adjustments in incomes paid to individual recipients which are tied to the inflation rate, usually measured by the Consumer Price Index.

Cournot Competition: An oligopoly in which each firm believes that its rivals are committed to a certain level of production and that rivals will reduce their prices as needed to sell that amount.

Credit Rationing: Credit is rationed when no lender is willing to make a loan to a borrower or the amount lenders are willing to lend to borrowers is limited, even if the borrower is willing to pay more than other borrowers of comparable risk who are getting loans.

Cross Subsidization: The practice of charging higher prices to one group of consumers in order to subsidize lower prices for another group.

Crowding Out: The tendency for federal government, by deficit financing to compete with firms or persons for borrowed funds; that is, firms and households unable to borrow at a low rate of interest curtail their investment and consumption spending.

Cryptosporidium: A protozoan (single-celled organism) that can infect humans, usually as a result of exposure to contaminated drinking water.

Cyclical Unemployment: Temporary layoff of workers due to downturns in the pace of economic activity.

Damage Function: Relationship that shows how pollution damage varies with the level of pollution emitted, and what the monetary value of that damage is.

Deficit Spending: A term which refers to the situation wherein he government spends more than it receives in taxes.

Demand: The maximum quantities of some good that people will choose (or buy) at different prices. An identical definition is the relative value of the marginal unit of some good when different quantities of that good are available.

Demand Curve: A graphic representation of the relationship between prices and the corresponding quantities demanded per time period. The relationship between quantity demanded of a good and the price, whether for an individual or for the market (all individuals) as a whole.

Demand Deposits: Checking accounts in commercial banks. These banks are obliged to pay out funds when depositors write checks on those numbers. Checking accounts are not cash – they are numbers recorded in banks.

Demand-Pull Inflation: A term used when an increase in aggregate demand occurs which cannot be offset by a corresponding increase in real supply causing an increase in the price level (inflation).

Demand Site Management: An attempt by utilities to reduce customers' demand for electricity or energy by encouraging efficiency.

Demographic Effects: Effects that arise from changes in characteristics of the population such as age, birthrates, and location.

Deposit/Refund Systems: A surcharge paid when buying potentially polluting products is refunded when the product or container is returned for recycling or proper disposal. Examples include "Bottle bills" deposits on beverage bottles and cans, containerized hazardous or solid waste, such as motor vehicle batteries, oil, and tires, and deposit-refund systems for car batteries. Recycling and environmentally safe disposal increase because the user is "paid" for doing it right.

Deregulation: The lifting of government regulations to allow the market to function more freely.

Developed Countries: The wealthiest nations in the world, including Western Europe, the United States, Canada, Japan, Australia, and New Zealand.

Dichotomous Choice: Offers respondents to a contingent valuation survey specific dollars and cents choices, for example would you be willing to pay between $10 and $20 per year to improve visibility at the Grand Canyon. Generally these amounts are varied between participants.

Diminishing Marginal Utility: The principle that says that as an individual consumes more and more of a good, each successive unit increases her utility, or enjoyment, less and less.

Diminishing Relative Value: The principle that if all other factors remain constant, and individuals relative value of a good will decline as more of that good is obtained. Accordingly, the relative value of a good will increase, other factors remaining constant, as an individual gives up more of that good.

Diminishing Returns: The principle that says as one input increases, with other inputs fixed, the resulting increase in output tends to be smaller and smaller as more and more of a productive resource is added to a given amount to other productive resources, additions to output will eventually diminish other factors, such as technology and the degree of specialization remaining constant.

Diminishing Returns to Scale: When all inputs are increased by a certain proportion, output increases by a similar proportion.

Dioxin: A man-made chemical by-product formed during the manufacturing of other chemicals and during incineration. Studies show that dioxin is the most potent animal carcinogen ever tested, as well as the cause of severe weight loss, liver problems, kidney problems, birth defects, and death.

Discount Rate: Degree to which future dollars are discounted relative to current dollars. Economic analysis generally assumes that a given unit of benefit or cost matters more if it is experienced now that if it occurs in the future. The degree to which the importance that is attached to gains and losses in the future is known as discounted. The present is more important due to impatience, uncertainty, and the productivity of capital. The interest a private bank pays for a loan from the U.S. Federal Reserve System.

Disequilibrium: The quantity demanded does not equal the quantity supplied at the going price.

Disinflation: A slowdown in the rate of inflation.

Disposable Income: The amount of an individuals income that remains after the deduction of income taxes.

Dividends: Profits of a firm that are distributed to its investors (stockholders).

Division of Labor: Assigning of specific tasks to workers and productive resources; it is a reflection of economic specialization.

Durable Goods: Goods that provide a service over a number of years, such as cars, major appliances, and furniture.

Economic Growth: A sustained increase in total output or output per person for an economy over a long period of time.

Economic Regulations: The control of entry into the market, pricing, the extension of service by established firms and issues of quality control.

Economic Rents: Payments made to a factor that are in excess of what is required to elicit the supply of that factor.

Economic Specialization: Concentration of activity in a few particular tasks or in producing only a few items.

Economics: The study of choice and decision-making in a world with limited resources.

Economies of Scope: What exists when it is less expensive to produce two products together than it would be to produce each one separately.

Efficiency: The allocation of goods to their uses of highest relative value.

Efficiency Wage: The wage at which total labor costs are minimized.

Efficiency Wage Theory: The theory that paying higher wages (up to a point) lowers total production costs, for instance by leading to a more productive labor force.

Efficient Markets Theory: The theory that all available information is reflected in the current price of an asset.

Effluent Fee: A fixed tax rate per unit (litre or kilogram) of emissions. They are also referred to as emission charges or emission taxes.

Elasticity of Demand: The percentage change in the quantity demanded divided by the percentage change in price.

Emission Charges: A fixed tax rate per unit (litre or kilogram) of emissions.

Emission Taxes: A fixed tax rate per unit (litre or kilogram) of emissions.

Entitlements: Government transfer payments made to individuals having certain designated characteristics and circumstances, such as age or need.

Equilibrium: The amount of output supplied equals the amount demanded. At equilibrium, the market has neither a tendency to rise nor fall but clears at the existing price.

Equilibrium Price: The price at which the quantity supplied and the quantity demanded are equal to each other.

Equivalent Variation: The amount of money one would accept to forgo a benefit such as a price decrease or the amount of income one would pay to avoid a harm such as a price increase. Money required to leave an individual as well off as after the economic change. Amount an individual would be willing to accept to forgo the change, or willing to pay to avert the change.

Exchange: The voluntary transfer of rights to use goods.

Exchange Rate: The price of one currency in terms of another.

Exchange Value: The purchasing power of a unit of currency for goods and services in the marketplace.

Exclusion Principle: The owner of a private good may exclude others from use unless they pay.

Existence Value: Value from knowing environmental goods exist independent of use or option value. If we lose a species in the wild, such as the Bengal tiger, very few of us will have our welfare directly affected by not being able to see it, photograph it or hear it. That "use value" is very small. But many people will lose the option to do that in the future, should they care to. Economists call that "option value." Further, many people around the world derive some benefit just from knowing that Bengal tigers exist in the wild. That is "existence value.".

Externalities: A situation in which an individual or firm takes an action but does not bear all the costs (negative externality) or receive all the benefits(positive externality). Costs or benefits that fall on third parties.

Factor Demand: The amount of an input demanded by a firm, given the price of the input and the quantity of output being produced; an input will be demanded up to the point where the value of the input's marginal product equals the price of the input.

Fiscal Policy: Policies that affect the level of government expenditures and taxes. Those federal-government expenditure, tax and borrowing decisions that affect the level of national economic activity.

Fixed Costs: The costs resulting from fixed inputs, sometimes called overhead costs.

Fixed Inputs: Inputs that do not change depending on the quantity of output, at least over the short term. Inputs that cannot be changed over a given time interval.

Free Good: A good which is abundant and costless.

Free Rider: One who receives something without paying.

Free-Rider Problem: Problem that occurs when someone thinks he may be able to enjoy something without paying for it, and fails to contribute ever a portion of the cost.

Frictional Unemployment: Unemployment due to workers leaving old jobs and seeking new ones.

Gains of Exchange: The difference between the relative values of a good to the buyer and the seller. How this difference is divided between buyer and seller will depend upon the price of the good. Exchange will not occur unless both the buyer and the seller expect to receive some of this gain.

GNP Deflator: Measure of the percentage increase in the average price of products in GNP over a certain base year (now 1972) published by the Commerce Department.

Good: Anything that anyone wants. All options or alternatives are goods. Goods can be tangible or intangible.

Government Budget Constraint: Total government outlays (the sum of expenditures on goods and services, transfer payments and interest on debt) must equal total revenue (the sum of taxes and U.S. government loans).

Government Security: A contract of the government promising to pay a lender a fixed rate of interest per year and repay the original loan at a fixed future date. Government Transfer Payment: Outlays by the government for which no good or service is received in the current period.

Gross National Product (GNP): The total market value, in terms of current dollars, of all final goods and services produced in the U.S. in one year.

Hedonic Pricing Approach: Derives values by decomposing market prices into components encompassing environmental and other characteristics through studying property values, wages and other phenomena. The premise of the approach is that the value of an asset depend on the stream of benefits derived, including environmental amenities.

Hypothetical Bias: Difference in actual willingness to pay and willingness to pay revealed in a survey arising from the fact that in actual markets purchasers suffer real costs, while in surveys they do not.

Inelastic Demand: A term used when the percentage change in quantity demanded is smaller than the percentage change in price.

Indexation: Modifying contracts so that their dollar terms adjust to the inflation rate as measured in an index, such as the consumer price index.

Inflation: Increase in the overall level of prices over an extended period of time.

Interest: The annual earnings that are sacrificed when wealth is invested in a given asset or business. The interest sacrificed by investing in a given business is often called the cost of capital.

Intergenerational Equity: Fairness between generations.

Intrinsic Values: Value that resides 'in' something and that is unrelated to human beings altogether.

Inventory: A stock of goods or resources held by a buyer or seller in order to reduce the cost of exchange or production.

Investment Expenditures: Dollar expenditures by firms on capital goods (factories, office buildings and others structures, machinery and equipment, inventories and residential housing) used to produce other new goods and services.

Involuntary Unemployment: Potential workers able and willing to work at the existing market wage rate, are unable to find jobs.

Irreversibility: If an asset is not preserved it is likely to be eliminated with little or no chance of regeneration.

Joint Implementation: Method of achieving reductions in CO_2 emissions whereby rich countries (which will probably have made binding commitments to cut emissions) can get partial credit for emission reductions projects which are funded by them, but which are undertaken in poor countries. This is now referred to as the Clean Development Mechanism (CDM) to denote respect for developing countries' right to develop. This mechanism is an attempt to make a system of marketable permits more equitable.

Labor Intensive Methods: Use of low quantity of capital per worker.

Labor Productivity: The ratio of real output per unit of labour input; growth is measured by a higher ratio of outputs to inputs.

Law of Demand: People purchase more of any particular good or service as its relative price falls; they purchase less as its relative price rises.

Law of Supply: At higher relative prices, the quantity supplied of a good will increase; at lower relative prices, smaller quantities will be supplied.

Leisure: All uses of time in which ones labor services are not exchanged for money. The uses of everyone's time can be divided between employment and leisure.

Liabilities: The debts of a person or business.

Macroeconomics: The study of the sum total of economic activity, dealing with the issues of growth, inflation and unemployment and with national economic policies relating to these issues.

Malthusian Trap: The minimum subsistence level to which humans descend as a result of geometric population growth and arithmetic resource growth.

Marginal: The additional or extra quantity of something. If one drinks six sodas in a day, the marginal soda would be the sixth soda.

Marginal Cost: The increase in total costs as one more unit is produced.

Marginal Productivity: The additional output obtained by adding an additional unit of a productive resource, such as labour. More precisely, marginal productivity is the change in total output divided by the change in the amount of the productive resource employed. Marginal productivity = change in total output change in amount of productive resource.

Marginal Propensity to Consume (MPC): The percentage of new or added income that is consumed.

Marginal Propensity to Save (MPS): The percentage of new or added income that is saved.

Marginal Revenue: The addition to total revenue as one additional unit is produced and sold.

Marginal Tax Rate: The tax rate charged on the taxpayers last dollar earned; in a progressive tax system the marginal tax rate is always greater than the average tax rate.

Market: A network in which buyers and sellers interact to exchange goods and services for money.

Market Clearing Price: A price which rations the supply of a good among competing consumers so that the quantity of the good demanded is equal to the quantity supplied.

Market Economy: A decentralized system where many buyers and sellers interact.

Mass Balance Condition: The mass of all the inputs used to produce goods and services (output) must equal the mass of the resulting output(s) plus the mass of the wastes.

Microeconomics: The study of the individual parts of the economy, the household and the firm, how prices are determined and how prices determine the production, distribution and use of goods and services.

Minimum Wage: A wage below which employers may not legally pay employees for specific kinds of employment.

Monopolistic Competition: A market with a large number of firms selling similar but differentiated products with no significant barriers to entry.

Monopoly: A market with only one supplier.

Multiplier: The number of times new investment spending will be respent to produce a certain amount of new income.

NAMEA: The national accounting matrix including environmental accounts developed by the Netherlands and used in their national income accounting reports. It contains figures on environmental burdens related to economic activity as reflected in the national accounts.

Natural Monopoly: One producer supplying all of the market at lower costs than many producers could.

Natural Unemployment Rate: An economy's civilian unemployment rate when supply and demand for labor are equal. The natural rate is the percentage of the civilian labor force unemployed at one time or another during any given year multiplied by the average time people spend searching for jobs.

Need: A specific quantity of a specific good for which an individual would pay any price.

Net Worth: The difference between the assets and liabilities of a person or business.

New Classical Macroeconomics: See Supply-Side Economics.

No Regrets Strategy: A strategy in response to the threat of climate change which argues that energy-saving measures should be undertaken immediately to help reduce global warming and climate change. Even if the threat of climate change is not as pronounced as we now fear, the supporters of this strategy say would not need to be any regrets because we would have benefited from saving the energy.

Nominal GNP: GNP measured in current prices (see Real GNP).

Nominal Interest Rate: The cost inflicted by inflation eroding the value of stored dollars plus the forgone real interest rate; the opportunity cost of holding money.

Normative Economics: Analysis that contains value judgments, either implicitly or explicitly (see Economics or Positive Economics).

Oligopoly: A market structure with just a few firms controlling a high percentage of total sales.

Open Access Resource: An open access resource is one where it is impossible to control the access of individuals who want to use it. Common examples are a fishery, or (in the classic example of the tragedy of the commons) a common pasture.

Opportunity Cost: The highest-valued sacrifice needed to get a good or service.

Option: Anything that anyone wants. In economics, options (alternatives) are also called goods.

Option Value: Potential benefits of the environment not derived from actual use. This expresses the preference or willingness to pay for the preservation of an environment against some probability that the individual will make use of it at some later date. If we lose a species in the wild, such as the Bengal tiger, very few of us will have our welfare directly affected by not being able to see it, photograph it or hear it. That "use value" is very small. But many people will lose the option to do that in the future, should they care to. Economists call that "option value." Further, many people around the world derive some benefit just from knowing that Bengal tigers exist in the wild. That is "existence value.".

Organization of Petroleum Exporting Countries (OPEC): A group of nations that produce most of the worlds oil and control most of the worlds oil exports.

Pareto Optimum: Situation in which it is impossible to make any individual better off without making someone else worse off, where better off means more preferred and worse off means less preferred. Every competitive market equilibrium is a Pareto optimum and every Pareto optimum is a competitive equilibrium if a set of assumptions (*e.g.* perfect information, absence of externalities, etc.) holds true.

Personal Saving: The difference between household income (after taxes) and consumption expenditures.

Political Economy: Policies that emphasize the interaction between politics and economics and that have political and economic effects.

Polluter Pays Principal: Policies that emphasize the interaction between politics and economics and that have political and economic effects.

Pollution Fee or Tax: Charge for the amount of waste or pollution. Examples include the BTU tax that was an early casualty in the President's budget bill. Several European nations have air and water pollution charges; Unit pricing for trash pickup, charging by the amount of trash collected (or the size of the container). The charge makes it worthwhile for a producer to cut back, right up to the point where it begins to cost more to reduce pollution than to pay the tax. A system like this also raises money for government, allowing government, if it chooses, to reduce taxes in other areas while collecting the same amount of total revenue.

Potential Pareto Improvement Criterion: The policy objective that gainers from a policy change (or project) *could* compensate the losers from the change and still be better off. In particular note that a policy that passes this criterion does not need to include the compensation, the compensation merely has to be possible.

Present Value: Value today of a sum to be paid or collected in the future to buy a good or service.

Price: The amount of money, or other goods, that you have to give up to buy a good or service.

Price Ceiling: The upper legal limit on a price.

Price Elasticity of Demand: A measure of the responsiveness of the quantity demanded of a good to changes in that goods price.

Price Floor: The lower limit imposed on a products price by a price control law.

Private Good: A good exclusively owned that cannot be simultaneously used by others. A good which when consumed by an individual is unavailable for others to consume.

Production Possibilities Curve: All combinations of the maximum amounts of goods that a society can produce with the available resources and technology.

Productive Resources: The inputs of labour, natural resources and capital used to generate new goods and services.

Profits: The excess of income over all costs, including the interest cost of the wealth invested. The net income of a business is not an accurate measure of its profit.

Property Rights: The conditions of ownership of an asset, the rights to own, use and sell. The right to use or consume something, *or* trade the right away in return for something else.

Prospect Theory: States that individual values with respect to gains and losses are in comparison to a reference point. Derived from psychology helps explain some anomalies including differences with respect to willingness to pay and willingness to accept. This contrasts with the economic assumption that individuals maximize utility. What matters is the point from which gains and losses are measured. It also suggests that values for negative deviations from the reference point will be greater than values place on positive deviations. Gains are valued less than losses. Third, the manner in which the gains and losses are to be secured matters a great deal.

Public Goods: Goods that cannot be withheld from people even if they don't pay for them. A good which, if made available to one person, automatically becomes available to all others in the same amount.

Pure Competition: A situation where many sellers sell the same product and no seller can set the price.

Quota: A quantitative restriction on imports.

Rational Expectations: Market participants intuitively anticipate systemic policy actions and their consequences for the economy; thus, on average, private market forecasts are accurate and planned policy is ineffectual (see New Classical Macroeconomics).

RCRA (Resource Conservation and Recovery Act): U.S. federal law originally enacted by Congress in 1976 to prevent the creation of toxic waste dumps by setting standards for the management of hazardous waste.

Real GNP: The GNP of any year measured in the prices of a base year. Real GNP is nominal GNP adjusted for inflation.

Real Rate of Interest: The dollar interest rate corrected for inflation; equal to the nominal rate minus the inflation rate.

Real Wage: Ones wage adjusted for inflation.

Regressive Taxes: A greater portion of income is taken from those in lower levels than from those in upper income levels.

Rent Controls: Fixed limits on rents that can be charged to tenants by owners according to a legal restriction.

Retained Earnings: Business profits which are held by firms and not paid to the stockholders of the firm; the earnings are usually reinvested by the firms.

Resource Conservation and Recovery Act (RCRA): U.S. federal law originally enacted by Congress in 1976 to prevent the creation of toxic waste dumps by setting standards for the management of hazardous waste.

Revenues: Total gross earnings of a firm before subtracting costs.

Scarce Good: A good which people want more of and which is costly to obtain.

Scarcity Shortage: A term used when the quantity of a good demanded exceeds the quantity supplied at the existing price.

Shadow Prices: Unobserved hidden or implicit prices derived through inferences and such methods as Contingent Valuation and Hedonic Pricing. Reflect movements along efficient frontier and tradeoffs between attributes.

Short Run: The period during which some inputs are fixed and cannot be varied.

Social Costs: Private costs plus external costs.

Specialization: The act of producing more of a good than one consumes, the rest of that good being exchanged.

Stagflation: An economic condition characterized by simultaneous inflation, slow growth and high unemployment.

Starting Point Bias: Because survey interviewers suggest the first bid this can influence the respondents answer and cause the respondent to agree too readily with bids in the vicinity of the initial bid.

Strategic Bias: Causes survey results to differ from actual willingness to pay because individual have an incentive to not reveal the truth because they can secure a benefit in excess of the costs they have to pay. This arises from the free rider problem. For example, if individuals are told that a service will be provided if the total sum they are willing to pay exceeds the cost of provision and that each will be charged a price according to their maximum willingness to pay then individuals will have an incentive to understate his or her demand.

Structural Unemployment: Workers without jobs whose skills are no longer suitable for or do not match the types of jobs available.

Substitutes: Price change for one product leads to a shift in the same direction in the demand for another product.

Supply Curve: A graphic representation of the relationship between quantities supplied at each price for a given time period.

Superfund Law: Comprehensive Environmental Response, Compensation and Liability Act (CERCLA).U.S. federal law enacted by Congress in 1980 for the purpose of cleaning up existing toxic sites.

Supply-Side Economics: Focus on the effects of national output potential or supply through reduction of taxes and government regulation for businesses designed to increase productivity and economic growth.

Surplus: A term used when the quantity of a good supplied exceeds the quantity demanded at the existing price.

Sustainable Development: A principle which states that a development plan must not compromise the welfare of future generations for the benefit of present generations.

Taking: Argument that government regulations can effectively *take* away or reduce the right of individuals or firms to use property to maximize their incomes or utilities.

Tariff: A tax on imports.

Technological Change: An advance, usually scientific, that causes an increase in output to occur relative to the quantity of inputs.

Terms of Trade: The relative prices of goods and services traded in international markets.

Trade-Off: The opportunity costs of selecting one alternative rather than another.

Tradeable Permits: The government specifies an overall level of pollution we'll tolerate, then gives each polluter a "permit" for its portion of the total. Firms that keep emissions below their allotted level may sell or lease the surplus to other firms that can use the permits to exceed their original allotment. For example, The 1990 Clean Air Act which set up tradeable permits for sulfur dioxide emissions in an effort to reduce acid rain. The approach may save the economy $1 billion a year. Other cases where it can work include water pollution from both point and non-point sources and international trading in greenhouse gas permits. If the number of permit holders is very high, the program can be expensive to operate. If the number is very small, some firms could monopolize the market.

Tragedy of the Commons: The case of a communal pasture area where all individuals are free to graze their livestock. The 'tragedy' arises because these 'commons' were typically heavily over grazed.

Travel Cost Method: Derives values by evaluating expenditures of recreators. Travel costs are used s a proxy for price in deriving demand curves for the recreation site.

Transaction Costs: The full costs of making an exchange.

Trough: A point in the business cycle corresponding to the end of the slowdown and the beginning of expansion.

User Benefits: Benefits deriving from the actual use of the environment. Anglers, hunters, boaters, nature walkers, bird watchers, etc. use the environment and derive benefits.

User Values: Benefits deriving from the actual use of the environment. Anglers, hunters, boaters, nature walkers, bird watchers, etc. use the environment and derive benefits. If we lose a species in the wild, such as the Bengal tiger, very few of us will have our welfare directly affected by not being able to see it, photograph it or hear it. That "use value" is very small. But many people will lose the option to do that in the future, should they care to. Economists call that "option value." Further, many people around the world derive some benefit just from knowing that Bengal tigers exist in the wild. That is "existence value.".

Variable Costs: Costs of a production process that increase or decrease along with changes in level of production, as opposed to fixed costs. Voluntary Export Restraint: Identical to an import quota except that the foreign market agrees voluntarily to limit exports from its county to a market.

Vehicle Bias: Difference in actual willingness to pay and willingness to pay revealed in a survey arising from the choice of a payment instrument for a survey. Vehicles include changes in local taxes, entrance fees, surcharges on bills, higher prices, etc.

Waste: When the relative value of a good is different from that goods marginal cost of production, waste occurs. Goods or resources are wasted when they are allocated to uses which are not the most valuable.

Wealth: The value of the existing stock of goods; those goods may be tangible or intangible.

Wholesale Price Index: A measure of changes in the prices of goods at the wholesale level, particularly those goods sold between businesses.

Willingness To Accept (WTA): Minimum amount of money one would accept to forgo some good or to bear some harm.

Working Poor: Workers earning inadequate income as judged by government-established standards of poverty.

WTA (Willingness To Accept): Minimum amount of money one would accept to forgo some good or to bear some harm.

WTP (Willingness To Pay): Maximum amount of money one would give up to buy some good.

Dictionary Links

Economics	Environment	General and Miscellaneous	Real Estate	Statistics
AmosWeb Economic Glossary	Appraisal Institute Glossary of Detrimental Conditions	Babylon.com Featuring Many Glossaries, Translation Info	Appraisal Institute Glossary of Detrimental Conditions	Hyperstat OnLine
Economics Dictionary The Dismal Scientist	National Park Service Ecology and Restoration Glossary includes Plant and Ecology Glossary; Wetlands Restoration Glossary	Dictionary.com	Bankrate.com Dictionary of Real Estate Terms	Norman Marsh's Glossary of Statistics
Joseph Stiglitz' Economics Dictionary from W.W. Norton	National Council for Science and the Environment (NCSE) formerly Committee for the National Institute for the Environment (CNIE) Dictionaries and Glossaries	Findlaw Search the Merriam Webster Legal Dictionary	Comps.com Guides	Oswego U. Online Glossaries for Statistics and Econometrics (Links)
Randy Wiggles' Glossary of Environmental Economics Terms	Randy Wiggles' Glossary of Environmental Economics Terms	Wordfocus.com Valuable Word Links	Freddie Mac Glossary of Terms	Statpoint.com Internet Statistics Directory
FACS Journalist's Guide to Economic Terms (broken)	Resources for the Future Glossary		RealtyGuides Links to Glossaries	Statsoft Statistics Glossary
United States Environmental Protection Agency Terms of Environment			Steps Statistics Glossary	
	Natural Resources Defense Council Environmental Terms, Laws and Treaties (broken)			

Other Educational Resources

American Agricultural Economics Association	Association of Environmental and Resource Economists	Dasgupta The Economics of the Environment
EconEdLink	Environmental Damage Valuation & Cost Benefit Website	Environmental Damage Valuation & Cost Benefit News
Environmental Damage Valuation & Cost Benefit Links	Environmental Protection Agency Economy and Environment Division	Committee for the National Institute for the Environment CNIE General Reference Bookmarks
Online Economics Textbooks	A Pedestrian's Guide to the Economy by Orly Amos, Professor of Economics, Oklahoma State U.	Resources for Economists RFE Teaching Resources
Resources for the Future	Robert Stavins : Can Market Forces Protect the Environment? From FACSNET	Robert Stavins Economic Coverage of the Environment from FACSNET
World Bank Environmental Economics and Indicators		World Resources Institute
Environmental Valuation & Cost Benefit Website Educational Kids Links		Environmental Valuation & Cost Benefit Website Economics Educational Kids Links

Index

A

Acid deposition 213
Acid formation 221
Acid rain 210, 215, 216, 219
Acidification 212
Actions 252
Adverse effects 214, 215
Aerosol formation 213
Amenity value 144
Animals 226
Appraisal method 138
Aquatic life 226
Aquatic systems 223
Arthropods 223
Assignment problem 46
Asymmetric information 60
Atmosphere 221

B

Basic model 51
Benefit approaches 169
Benefit transfer 149
Benefit transfer method 125, 126, 150

Biodiversity 183, 184, 232, 244, 260
Biodiversity loss 190, 192
Biofuels 180
Birds 226

C

Carrying capacity 26
Chemistry of corrosion 220
Circular flow 10
Climate change 183, 229
Climate models 201
Climate variations 200
Cloud Droplets 213
Club goods 17
Coasian solutions 46
Coastal areas 254
Coastal environment 9
Cobweb model 69
Common property 19
Common resources 17
Conserving environmental resources 257
Contingent choice method 109, 110
Contingent valuation 147

Contingent valuation method 102, 104, 148

Coral reefs 180

Corrective 55

Corrective taxation 49, 55

Cost approaches 169

Cost benefit calculations 170

Cost estimates 207

Cost–benefit analysis 165

Costs of reduction 56

Criminal development 86

Cross-cutting methods 149

D

Damage 221

Damage cost avoided, replacement cost 130

Decentralisation 251

Deforestation 244

Desertification 245

Development 182, 204

Direct effects 205

Dirty subsidies 21

Discount 186

Divisibility 67

Dose-illness models 161

Dose-infection models 161

Dose-response modelling 155

Dry deposition 214

Dynamic government failure 72

Dynamic market failure 69

E

Ecocentrism 86

Ecological economics 152

Ecological footprint 27

Ecological resource valuation 153

Ecological valuation 152

Econometric considerations 83

Econometric framework 77

Economic distortions 71

Economic instruments 237

Economic view 7

Economics 284

Economics and the environment 4

Economics of biodiversity 179

Economy 10

Ecosystem valuation 30

Emissions trading 216

Enforcement 255

Engineering 4

ENV 2A8Y 2

Environment 284

Environment kuznets curve 74

Environment policy consumption 236

Environmental 205

Environmental damages 236

Environmental economic Values 89

Environmental economics 1, 24

Environmental externalities 38, 39, 42

Environmental goods economic evaluation 90

Environmental goods monetary evaluation 88

Environmental governance 249

Environmental impact evaluation techniques 165

Environmental offsetting 252

Environmental regulations 20

Environmental resource damages 91

Environmental resource use 249

Environmental resources 249

Environmental standard setting 251

Environmental taxes 239

Environmental valuation 5, 86

Environmental values 89

Environmentally sensitive zones 255

Equity 71, 250

Ethics 186

Evaluation challenge 191

Excludability 67

Exhaustible resources 10

Existence value 96

Explanations 74

Exposure 156

Externalities 37, 49, 61, 64

Extrapolation 162

F

Factor income method 146

Farmed oysters 14

Financing sustainable development 242

Fisheries 185

Flue gas desulfurization 215

Forest floor 216

Forests 258

Fossil fuel 206

Free rider problem 48

Freshwater resources 261

Functional forms 78

G

Gas phase chemistry 213

GDP 189

Gender 184

Global environmental issues 229

Global warming 196, 202, 242

Government failure 70

Green accounting 238

Green movement 240

Greenhouse effect 197, 229

Gross primary energy valuation 152

Groundwater 262

Growth 182, 224

H

Hazardous waste 243

Health 205

Hedonic price method 143

Hedonic pricing method 127

Holdout problem 47

Human beings 226

Human carrying capacity 27

Human emissions 212

Human-made materials 218

Hydrolysis 213

Hypothetical scenario 109

Hypothetical situation 116, 123, 125, 128, 131

I

Illness 157

Incidental value 94

Independent action 160

Indirect use values 92

Infection 157

Infection-illness models 161

Information 68, 71

Infrastructure 203

Insecurity 42

Integration 251

Intergenerational equity 187

International treaties 216

Intra-generational equity 249

Investment 203

K

Key concepts 2

Key issues 7

Key terms 180

Kuznetsian relationship 82

L

Lagged adjustment 69

Lake ecology 214

Land degradation 257

Law of demand 28

Law of diminishing returns 24

Law of supply 28
Legal action 253
Legal liability 250
Legislative framework 252
Living modified organisms (LMOs) 255
Low dose extrapolation 162

M

Marginal social benefit 64
Marginal social cost 64
Marine pollution 243
Market allocation 12
Market failure analysis 59
Market failures 15, 38
Market power 61
Market price approach 137
Market price method 98, 100
Market-based techniques 136
Marketed natural resources 166
Meaning 37
Migration 203
Mitigation 207, 208
Modelling concepts 158
Monitoring 255
Mortality 158, 162

N

Natural emissions 212
Natural heritage 260
Natural resource accounting 236
Natural resource management 4
Natural resources 12
Negative consumption 41
Negative externalities 44
Negative production 39
NEP 2004 248
Net present value 29
Non market valuation 140
Non-marketed environmental effects 167

Non-threshold mechanisms 159
Non-use value 93
Northwest passage 204

O

Optimist's paradox 187
Oxidation 213
Ozone depletion 231
Ozone layer depletion 243

P

Pathogen-host-matrix triangle 162
Policy failures 38
Policy response 230, 232, 233
Pollution 21
Positive externalities 43
Poverty 184
Prevention methods 215
Preventive action 251
Price and quantity 57
Price option 95
Price regulation 55
Private-sector 44
Process related reforms 253
Production function approaches 166
Productivity method 122
Property rights 21, 44, 237
Public goods 19
Public trust doctrine 251
Public-sector remedies 49

Q

Quasi option value 95

R

Random utility modeling 142
Real estate 284
Reduction costs 53
Regression equations 78
Regulation 50
Regulatory failure 62

Regulatory reforms 252
Renewable resources 10
Reproductive failure 224
Resource replacement cost method 139
River systems 261

S

Science 4
Selection of models 160
Sequelae 158
Side effects 70
Skeletal deformity 225
Social sciences 4
Soil biology 215
Solar variation 198
Solutions 227
Sources 37, 225
Statistics 283
Strategies 252
Subsidies 50, 185
Substantive reforms 253
Sustainability 22
Sustainable development 234
SUVs 42
Synergistic action 160

T

Tariffs 21
Tax factors 171
Taxes 21
Technical solutions 215
Technocentrism 86
Temperature 205

Temperature changes 199
Theoretical background 75
Three pillars 22
Time preference 154
Total economic value 87, 91
Tradable permits 55
Trade 238
Traditional knowledge 260
Transaction costs 48
Transboundary movements 243
Travel cost method 115, 120, 141
Travel cost model 142
Types of market failure 60

U

Unit day value method 150
Unresolved issues 154

V

Valuation 20
Valuation of effects 168
Valuation results 194
Value of life 144
Vicarious consumption value 94
Vicarious value 97

W

Water 217
Water scarcity 205
Wet deposition 213
Wildlife 258, 259

Z

Zonal travel cost approach 117

www.ingramcontent.com/pod-product-compliance
Lightning Source LLC
Chambersburg PA
CBHW020217290326
41948CB00001B/79